INSECTICIDES - AGRICULTURE AND TOXICOLOGY

Edited by **Ghousia Begum**

Insecticides - Agriculture and Toxicology
http://dx.doi.org/10.5772/intechopen.68177
Edited by Ghousia Begum

Contributors

Partida Ruvalcaba Leopoldo, Medina López Raymundo, Cruz Ortega Jacobo Enrique, Díaz Valdés Tomás, Velázquez Alcaraz Teresa De Jesús, Ayala Tafoya Felipe, Edgar Antonio Reyes-Montaño, Nohora Angélica Vega-Castro, Teodoro Stadler, Micaela Buteler, Javier Gustavo Gitto, Susana Ruth Valdez, Shaon Kumar Das, Irani Mukherjee, Karina Buzzetti, Maria Vagi, Andreas Petsas, Maria Dimitriou Pavlaki, Maria Kostopoulou, Niki Smaragdaki, Melina Espanhol Soares, Flávio Soares Silva

First published in London, United Kingdom, 2018 by IntechOpen
IntechOpen is the global imprint of INTECHOPEN LIMITED, registered in England and Wales, registration number: 11086078, The Shard, 25th floor, 32 London Bridge Street
London, SE19SG – United Kingdom
Printed in Croatia

British Library Cataloguing-in-Publication Data
A catalogue record for this book is available from the British Library

Additional hard copies can be obtained from orders@intechopen.com

Insecticides - Agriculture and Toxicology, Edited by Ghousia Begum
p. cm.
Print ISBN 978-1-78923-166-3
Online ISBN 978-1-78923-167-0

For EU product safety concerns :
IN TECH d . o . o . , Prolaz Marije Krucifikse Kozulić 3 , 51000 Rijeka , Croatia ,
info@intechopen . com or visit our website at intechopen . com .

Meet the editor

Dr. Ghousia Begum got her PhD degree from Osmania University and continued her postdoctoral work in the same University. In 1998, she was appointed as junior scientist in Applied Biology Division, CSIR-Indian Institute of Chemical Technology. She has significantly contributed to the field of *ecotoxicology, biochemical and molecular mechanisms of toxicity, and environmental biology*. Her research interests lie on how environmental changes, particularly toxicants, affect physiological functions in aquatic animals especially in fish. She is also working on *alternative animal models, which include lower invertebrates (Daphnia) and vertebrates (zebra fish and edible fishes) for toxicity evaluation*. She has published more than 30 research articles in reputed journals. She has more than 500 citations to her credit.
She has been awarded Young Scientist Award for the year 1993 by the National Environmental Science Academy (NESA), gold medal for academic excellence, and Scientist of the Year Award (2012) by NESA. She has edited a book entitled *Ecotoxicology* published by InTech. Presently, she is an editorial board member in eleven journals and an editor of two journals. She has worked as a peer reviewer with many journals in the area of environmental biology, toxicology, and related areas.

Contents

Preface

Insecticides are substances used to kill insects. They are used primarily in agriculture to control pests that infest crop. Nearly all insecticides have the potential to significantly alter ecosystems: many are toxic to humans and/or animals; some become concentrated as they spread along the food chain. The presence of these chemicals in both aquatic and terrestrial ecosystems has become an important issue globally. The book *Insecticides-Agriculture and Toxicology* provides information on the use of insecticides in pest management in order to enhance crop protection and their effects on nontarget organisms. The results of the efficacy of biorational insecticides on diatomaceous earth for the control of leaf miner in chickpea and Mexican bean weevil are reported; the chapter concluded with recommended dose of diatomaceous earth per kilogram of seed. The next chapter is from Protein Research Group on isolation and characterization of lectins from Colombian seeds of Fabaceae and Lamiaceae plants. They have shown that lectins have a high potential as insecticide or insectistatic agents. Information on the impact of active ingredient on pest study on different commercial formulation of diazinon, acetamiprid, lambda-cyhalothrin, and imidacloprid insecticides in the control of important pests of apple is described in detail, and data are expressed clearly. One chapter is devoted to various health problems caused by insecticide exposure, and the use of protective clothing and equipment is obligatory from the manufacturing stage to the final application on pests. One chapter is on nanobased innovative nanoinsecticides, which have broad-spectrum pest protection efficiency, reducing water, soils, and environmental pollution in comparison with conventional insecticides. This chapter deals with the mode of action of nanostructured alumina and demonstrated that interaction of nanostructured alumina with insects' cuticle is the main insecticidal efficacy. Propesticides are important agrochemicals with optimal efficacy and environmental safety. One chapter gives sufficient space to describe comprehensively the successful utilization of propesticides and their activation processes and classification based on the type of pests. The last chapter is on acute toxicity of commonly used organophosphate insecticide, fenthion, to aquatic microorganism and marine algae. The effects of fenthion on chlorophyll pigments were observed, which can be used as a biomarker of toxicity. The quantitative structure activity relationships (QSARs) were applied to compare observed and predicted toxicity results.

I hope that the components of this book will suffice the requirements of the researchers, scientists, and students from agriculture, agrochemicals, toxicology, aquatic toxicology, ecology, and other related areas.

Dr. Ghousia Begum
CSIR-Indian Institute of Chemical Technology
India

Chapter 1

Biorational Insecticides and Diatomaceous Earth for Control Sustainability of Pest in Chickpea and Mexican Bean Weevil

Jacobo Enrique Cruz Ortega,
Leopoldo Partida Ruvalcaba,
Raymundo Medina López, Tomás Díaz Valdés,
Teresa de Jesús Velázquez Alcaraz and
Felipe Ayala Tafoya

Additional information is available at the end of the chapter

http://dx.doi.org/10.5772/intechopen.71534

Abstract

Sustainability involving the conservation and/or enhancement of natural resources and environmental protection can be practiced with biorational insecticides or diatomaceous earth. Two researches were carried out; in one, the objective was to determine the efficacy of biorational insecticides in controlling chickpea leaf miner, *Liriomyza sativae* Blanchard, without completely inhibiting the presence of parasitoids of this pest. Biorational insecticides were chlorantraniliprole, cyromazine and spinosad, and conventional insecticide was chlorpyrifos, which were similarly effective to control adults and larvae of Liriomyza. Most chickpea production in 2012–2013 (1993.3 and 1806.8 kg ha^{-1}) was obtained where chlorantraniliprole and chlorpyrifos were applied, respectively, and where spinosad and cyromazine were applied also exceeded the performance of absolute control (1213.6 kg ha^{-1}). In 2013–2014, the increased production was 1621.9 kg ha^{-1} with chlorantraniliprole and 1556.3 kg ha^{-1} with chlorpyrifos, significantly different from the absolute control that produced 1136.5 kg ha^{-1}. Earnings were MX$ 21011.7 in 2012–2013 and MX$ 16036.7 in 2013–2014 with chlorantraniliprole, while in the absolute control, earnings were MX$ 12305.1 and MX$ 11083.5. Chlorantraniliprole was the biorational insecticide that caused greater effect in the management of this pest of chickpea and crop yields. While in another research, the objective was to determine the efficacy of different doses of diatomaceous earth against Mexican bean weevil *Zabrotes subfasciatus* Boheman. An experiment was carried out in two phases: in first, one tested diatomaceous earth at doses of 1.0, 2.0, 3.0, 4.0, and 5.0 g kg^{-1} of seed, with samples at 15, 30, 45, and 60 days after application (daa), while in the second, the doses were 0.2, 0.4, 0.6, 0.8, and 1.0 g kg^{-1} and samples at 10, 20, 30, and 40 daa. The parameters evaluated were weevil mortality and seed germination. The results indicated that the doses from 0.8 to 5.0 g kg^{-1} of diatomaceous earth efficiently controlled the Mexican bean weevil. The treatments did not inhibit seed germination.

IntechOpen

Keywords: *Liriomyza sativae* Blanchard, parasitoids, *Zabrotes subfasciatus* Boheman, chickpea, beans

1. Introduction

The chickpea crop is the second most important grain of the family Fabaceae grown in Asia, Mediterranean regions, Australia, Canada, USA, and Africa [1]. Globally, the chickpea is planted on an area of nearly 12 million hectares with an approximate production of 11,308,684 tons. This crop develops during the winter under different agroclimatic conditions; its production in Mexico is 271,894 tons annually, of which Sinaloa and Sonora generate 70 and 20%, respectively. Most of it is destined for the international market [2]. Chickpea is an important crop for Sinaloa due to the harvested area, volume, and quality of grain produced over an area of 60,000 hectares, with an average yield of 1.7 tons per hectare [3].

The study of the biological activity of some compounds present in plants offers an opportunity to discover new and efficient insecticides for pest control [4, 5] which could be tolerated by crops and harmless to consumer; different researches have been observed and reported the insecticidal action of different plant extracts [6–8]. The objective of this research was to determine the efficacy of biorational insecticides to sustainably control the leaf miner (*Liriomyza sativae* Blanchard) without totally inhibiting the presence of the parasitoids of this pest in chickpea cultivation.

On the other hand, beans is one of the most sown and consumed legumes in Mexico. During the 2014–2015 agricultural cycle, 220,263 ha was sown at the state of Sinaloa, 58,550 ha, to be placed in the first place in terms of sowing and harvesting of this grain [9]. Bean is one of the essential foods for the world population, which makes necessary the conservation and protection of this grain against various factors that affect it, within these stands the importance of the Mexican weevil (*Zabrotes subfasciatus* Boheman, Coleoptera: Bruchidae) of the bean. The larva feeds on grain and causes severe damage and decreases the germinative power of the seed, by considerably damaging the cotyledons, on which the damages by oviposition and the perforations that are the feeding chambers of these insects can be observed [10, 11].

Storage pests are one of the most important problems in storage of grains and seeds, if they are not controlled in a timely manner, and cause direct damages that affect the conservation of the grain. Likewise, they cause damages indirectly when they are invaded by various microorganisms such as fungi and bacteria that contaminate them, which can cause problems in humans when consumed. There are few products that can be used reliably for stored grains for pest control, mainly insecticides and fumigants that are not very persistent in Mexico and in the world, which makes it necessary to seek more alternatives to reduce the damages that cause the pests of stored grains, which do not affect the environment and human health. At the global level, different alternatives for the control of storage pests have been tested. These include treatments based on heat and cold; the use of plant extracts and mineral substances; pheromones; biological control; and the use of chemicals that are preferably under residual power and do not cause effects on grains, seeds, and consumers [12–15].

2. Biorational insecticides for control sustainability of leaf miner (*Liriomyza sativae* Blanchard) in chickpea (*Cicer arietinum* L.)

The research on *Liriomyza sativae* Blanchard was performed by two experiments that were established in the experimental field of the Faculty of Agronomy of the Autonomous University of Sinaloa, located at 17.5 km of Culiacan-Eldorado road, Culiacan, Sinaloa, Mexico, with coordinates 24° 48′ 30″ N, 107° 24′ 30″ W and 38.54 m. The climate of this region is very warm to semidry. Average annual rainfall varies from 500 to 700 mm. The average annual maximum temperature is 25°C. The soils of this region are predominantly clayey [16].

The experiment design was randomized complete blocks with four replicates, where the experimental plot consisted of six furrows of 10 m long with 0.8 m distance from each other. The useful plot was the two central grooves minus 1 m from each end. The first planting took place on December 21, 2012 and the second planting on December 30, 2013, both manually with a density of 15 plants per linear meter. Five treatments were evaluated: three biorational insecticides: chlorantraniliprole + ethoxylated alkyl aryl phosphate ester (100 mL + 1.0 L ha^{-1}), cyromazine + *Bacillus thuringiensis* (80 g + 1 kg ha^{-1}), spinosad + sugar (416.6 mL + 2.08 kg ha^{-1}), cyromazine + *Bacillus thuringiensis* (80 g + 1 kg ha^{-1}), spinosad + sugar (416.6 mL + 2.08 kg ha^{-1}); one conventional insecticide chlorpyrifos + ester ethoxylate alkyl aryl phosphate (1.5 L + 1.0 L ha^{-1}); and absolute control (without application of insecticides), applying them on the foliage twice. This was done in each evaluation year.

Two applications per cycle were performed on February 9 and March 16, 2013; 02 and 23 February, 2014 with a Maruyama motor pump with a capacity of 25 L, an output boom, and cone nozzle TX5, whose water expenditure was 208 L ha^{-1}. The insecticides were applied when the population and leaf miner damage exceeded the economic threshold of 20% to the foliage [17].

Samples of live larvae and empty mines were carried out weekly on a leaf of 10 randomly selected plants. Of each useful plot, 100 leaves were collected and confined in 0.5-L plastic containers at room temperature. After 12 days, the adult miners and emerged parasitoids were separated and confined in glass flasks with 70% alcohol. For identification of the miner, the male abdomen was introduced into a 10% potassium hydroxide solution to soak the tissue for 10 minutes at 80°C and then washed with distilled water to remove the potassium hydroxide. With the preparation immersed in 70% alcohol, the cuticle and tissues were separated from the abdomen until the complete genitalia were cleaned and exposed [18]. With the help of codes and schemes of the male genitalia published by Spencer and Stegmaier [19] and Spencer and Steyskal [20] the taxonomic determination was made. Identification of the parasitoids emerged from the leaf samples was carried out using the keys of Wharton [21] for the genus of the Braconidae family, whereas for the genus of the Eulophidae family, the keys of La Salle and Parrella [22].

To determine the percentage of damage, weekly damage and healthy leafs of three plants per repetition were counted, and the percentage of damage was calculated with a three rule simple modified. Harvest was performed when the culture reached its physiological maturity and the data were transformed to be analyzed with the statistical package SAS 9.1 [23] and then this showed significant differences that were submitted to Duncan's multiple range test with $\alpha = 0.05$ for mean separation.

While in the entomology laboratory of the same, faculty research was done to determine the efficacy of diatomaceous earth doses, where the colony of beans weevil (*Zabrotes subfasciatus* Boheman) was purified in glass bottles with a capacity of 5 kg, which were kept under a temperature that fluctuated between 30 and 35°C, with a purpose of having a homogeneous colony for the test.

To establish the tests; polystyrene beakers with capacity of 500 g, and 2 kg of bean per treatment were used; the application of the diatomaceous earth was homogenized on the grain, then 20 adults of bean weevil were deposited in each repetition and covered with organza cloth. The investigation was carried out in two phases: (a) the first one was done in a completely random design with seven treatments and four repetitions. The treatments were diatomaceous earth at doses of 1.0, 2.0, 3.0, 4.0, and 5.0 g kg^{-1} of seed, a chemical control (deltamethrin) at a dose of 1.0 mL kg^{-1} of seed, plus an absolute control (without application of substances); (b) the second phase consisted of another completely random experimental design with the same amount of treatments and repetitions, but with the doses of 0.2, 0.4, 0.6, 0.8, and 1.0 g kg^{-1} of diatomaceous earth, a chemical control (deltamethrin) at a dose of 0.1 mL kg^{-1} of seed plus the absolute control.

In the both phases of the experiment, the response variables were the percentage of dead adults and the germination of bean seeds. In the first phase, mortality was determined with the number of live and dead insects in each experimental unit, at 15, 30, 45, and 60 days after application (daa), while in the second phase, it was done at 10, 20 30, and 40 daa. With the averages of mortality in each experimental unit, the percentage of effectiveness was obtained by the following Eq. [24]:

$$\text{Corrected mortality} = \frac{(\text{mortality of the treatment–mortality of the absolute control}) \times 100}{100 - \text{mortality of the absolute control}}$$

Germination was evaluated with 100 bean seeds planted in polystyrene trays filled with peat moss and determined at 10, 20 and 30 daa of the diatomaceous earth and deltamethrin doses, counting the seedlings emerged in each of the experimental units and comparison of averages with respect at average of the absolute control, while percentages were also determined with the equation of Abbott [24]. All data were subjected to an analysis of variance and multiple comparison of means of Tukey test ($\alpha = 0.05$) of the statistical package SAS 9.1 [23].

3. Efficacy of biorational insecticides on *Liriomyza sativae* Blanchard without totally inhibiting the presence of the parasitoids

The leaf miner species present in the chickpea is *Liriomyza sativae* Blanchard. The aedeagus presents a barely conspicuous constriction (**Figure 1A**), where the edges of the distifalo have only a slight undulation. The ejaculatory pump apodema has a thin base that is wider at the distal end than the diameter of the bulb (**Figure 1C**).

http://dx.doi.org/10.5772/intechopen.71534

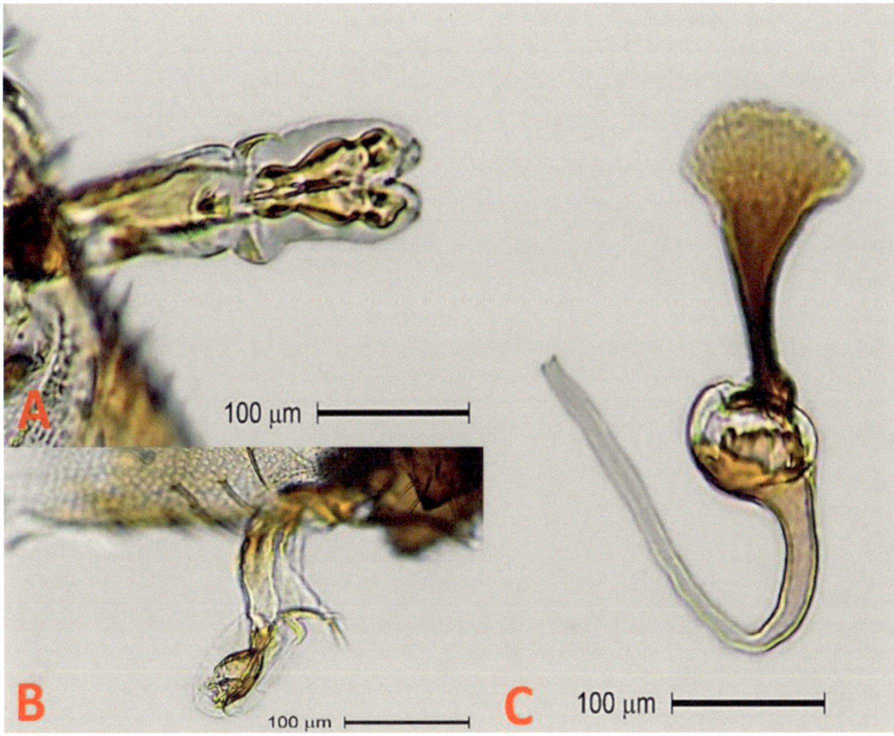

Figure 1. Liriomyza Sativae: (A) aedeagus ventral view, (B) aedeagus side view, and (C) sperm pump.

Based on the final averages of the two experiments (**Table 1**), it could be interpreted that after applying the biorational insecticides, chlorantraniliprole, cyromazine, and spinosad, twice as well as the conventional insecticide, chlorpyrifos, there was no statistical difference between the efficacy of the biorationals and that of the conventional one, and live larval populations of *L. sativae* were reduced to 43, 33, 22, and 39%, respectively, compared to the 100% presence of larvae in the foliage on average in the absolute control in the cycle 2013, while in 2014, the respective decreases were 55, 47, 45, and 46%. The tendency of the population of *L. sativae* to stay below the economic threshold in the two experiments, from the second application, was perhaps due to the physiological maturity of the chickpea approaching the senescence of the foliage and, consequently, to the harvest grain.

According to the final averages of **Table 2**, in 2013, the percentage of empty mines was reduced to 29, 37, 11, and 29% with chlorantraniliprole, cyromazine, spinosad, and chlorpyrifos, respectively, compared with 100% represented by the average of the control. In 2014, the respective decreases were 19, 3, 16, and 22% with the same treatments. In addition, the time of action of biorationals was very similar to that of chlorpyrifos. The results allowed to corroborate that the biorationals are products that can be used for the control of *L. sativae* with the same effectiveness of the conventional insecticide chlorpyrifos.

Treatments	February 2013				March 2013				Final Average
	7[1]	14	21	28	7	14[2]	22	28	
Chlorantraniliprole	2.30	1.65	0.20^{bc*}	0.35^{b}	1.65	3.95	0.40^{b}	0.05^{b}	1.17^{b}
Cyromazine	2.20	1.85	0.30^{b}	0.55^{b}	1.75	3.80	0.70^{ab}	0.75^{ab}	1.38^{b}
Spinosad	2.00	1.85	0.75^{a}	0.80^{ab}	1.90	4.15	0.40^{b}	1.30^{a}	1.59^{ab}
Chlorpyrifos	2.30	1.15	0.10^{c}	0.35^{b}	1.65	3.70	0.20^{b}	1.60^{a}	1.25^{b}
Absolute control	2.35	2.30	0.75^{a}	1.55^{a}	1.90	3.95	2.00^{a}	1.90^{a}	2.05^{a}
	February 2014				March 2014				
	7[3]	14	21[2]	28	7	14	22	28	
Chlorantraniliprole	0.70^{ab}	1.60	2.65^{b}	0.83^{b}	1.58	1.20	1.00	1.05^{b}	1.32^{b}
Cyromazine	0.78^{ab}	1.80	2.70^{b}	0.73^{b}	1.85	1.20	1.88	1.68^{b}	1.57^{b}
Spinosad	1.58^{ab}	1.83	2.75^{b}	0.85^{b}	1.35	1.98	1.40	1.25^{b}	1.62^{b}
Chlorpyrifos	0.75^{b}	1.85	3.85^{a}	1.00^{ab}	1.88	1.25	1.03	1.05^{b}	1.58^{b}
Absolute control	3.90^{a}	3.83	3.68^{a}	2.03^{a}	2.68	2.30	2.13	3.10^{a}	2.95^{a}

*Means with the same letter in each column are statistically the same (Duncan $\alpha \leq 0.05$).
[1]Two days before the first application.
[2]Two days before the second application.
[3]Five days after the first application.

Table 1. Average live larvae of leaf miner *Liriomyza sativae* Blanchard in 40 chickpea leaves in Culiacan, Sinaloa, Mexico.

Treatments	February 2013				March 2013				Final average
	7[1]	14	21	28	7	14[2]	22	28	
Chlorantraniliprole	1.58	1.73	1.03^{a*}	0.38^{ab}	0.83	0.65	1.33	0.98^{bc}	0.99^{bc}
Cyromazine	1.30	1.30	0.68^{ab}	0.60^{a}	0.78	1.00	1.15	0.70^{c}	0.88^{c}
Spinosad	1.38	1.58	0.80^{ab}	0.85^{a}	1.18	1.13	1.68	1.50^{a}	1.24^{ab}
Chlorpyrifos	1.28	1.70	0.38^{b}	0.08^{b}	1.30	1.33	1.23	0.95^{bc}	0.99^{bc}
Absolute control	1.38	1.78	1.38^{a}	0.85^{a}	1.33	1.38	1.95	1.15^{ab}	1.40^{a}
	February 2014				March 2014				
	7[3]	14	21[2]	28	7	14	22	28	
Chlorantraniliprole	2.58^{b}	0.28^{b}	1.08	1.68	1.10	3.20^{ab}	3.05	2.43^{ab}	1.92^{b}
Cyromazine	2.20^{b}	0.98^{ab}	1.45	1.38	0.83	3.93^{a}	3.88	3.68^{a}	2.29^{ab}
Spinosad	3.53^{a}	1.33^{a}	1.23	1.63	0.83	2.08^{b}	2.55	2.70^{ab}	1.98^{ab}
Chlorpyrifos	1.35^{c}	0.85^{ab}	1.50	1.13	1.10	3.75^{a}	2.85	2.25^{b}	1.84^{b}
Absolute control	2.78^{ab}	1.93^{a}	1.45	1.48	1.28	3.40^{ab}	3.28	3.43^{ab}	2.37^{a}

*Means with the same letter in each column are statistically the same (Duncan $\alpha \leq 0.05$).
[1]Two days before the first application.
[2]Two days before the second application.
[3]Five days after the first application.

Table 2. Average empty mines of leaf miner *Liriomyza sativae* Blanchard in 40 chickpea leaves in Culiacan, Sinaloa, Mexico.

Treatments	February 2013				March 2013				Final
	7[1]	14	21	28	7	14[2]	22	28	Average
Chlorantraniliprole	1.40^{ab}	1.49^{ab}	1.00^{ab}	0.77^{bc}	1.59^{ab}	1.61	1.25^{b}	0.79^{b}	1.21^{b}
Cyromazine	1.30^{ab}	1.05^{ab}	0.84^{b}	0.60^{c}	0.89^{b}	2.08	1.59^{ab}	1.40^{ab}	1.20^{b}
Spinosad	1.29^{ab}	1.73^{a}	1.52^{a}	1.39^{ab}	1.95^{ab}	2.16	2.50^{ab}	1.01^{ab}	1.75^{ab}
Chlorpyrifos	1.79^{a}	0.61^{b}	1.56^{a}	0.51^{c}	2.34^{a}	1.09	1.07^{b}	3.04^{a}	1.46^{ab}
Absolute control	0.98^{b}	1.51^{ab}	1.59^{a}	1.70^{a}	1.74^{ab}	1.85	3.15^{a}	2.36^{ab}	1.98^{a}
	February 2014				March 2014				
	7[3]	14	21[2]	28	7	14	22	28	
Chlorantraniliprole	1.18^{b}	2.05^{b}	0.35^{b}	0.73^{ab}	4.30	3.93	1.20^{b}	9.78	2.94^{bc}
Cyromazine	0.00^{c}	1.38^{b}	0.08^{b}	0.28^{b}	7.20	2.08	1.05^{b}	2.63	1.83^{c}
Spinosad	4.15^{a}	6.88^{a}	0.25^{b}	1.15^{ab}	10.93	2.75	5.08^{a}	7.78	4.87^{a}
Chlorpyrifos	0.35^{bc}	9.95^{a}	0.03^{b}	0.75^{ab}	8.70	0.38	5.28^{a}	6.85	4.03^{b}
Absolute control	5.98^{a}	6.53^{a}	1.13^{a}	1.68^{a}	9.98	2.93	7.10^{a}	7.85	5.39^{a}

*Means with the same letter in each column are statistically the same (Duncan $\alpha \leq 0.05$).
[1]Two days before the first application.
[2]Two days before the second application.
[3]Five days after the first application.

Table 3. Average adult leaf miner *Liriomyza sativae* Blanchard in 40 chickpea leaves in Culiacan, Sinaloa, Mexico.

Treatments	February 2013				March 2013				Final
	7[1]	14	21	28	7	14[2]	22	28	Average
Chlorantraniliprole	20.1	9.7^{b}	9.9^{b}	5.1^{b}	2.8^{b}	20.4	8.4^{b}	4.5^{b}	8.7^{b}
Cyromazine	20.2	9.1^{b}	5.9^{bc}	6.6^{b}	5.0^{b}	20.1	7.5^{b}	8.1^{b}	8.9^{b}
Spinosad	20.1	12.2^{b}	4.9^{c}	8.6^{b}	4.6^{b}	22.1	8.0^{b}	8.6^{b}	9.9^{b}
Chlorpyrifos	21.2	9.3^{b}	4.2^{c}	6.2^{b}	3.2^{b}	22.0	8.9^{b}	9.5^{b}	9.0^{b}
Absolute control	20.9	22.7^{a}	22.1^{a}	22.6^{a}	31.2^{a}	25.8	21.8^{a}	23.9^{a}	24.3^{a}
	February 2014				March 2014				
	7[3]	14	21[2]	28	7	14	22	28	
Chlorantraniliprole	11.2^{ab}	9.8^{bc}	20.2	13.9	9.7	8.9	9.4^{b}	7.9^{d}	11.4^{d}
Cyromazine	16.4^{ab}	14.8^{ab}	21.0	19.7	7.6	11.1	9.3^{b}	10.9^{cd}	13.8^{bc}
Spinosad	18.5^{a}	13.8^{ab}	21.3	16.8	6.7	14.9	12.4^{ab}	15.4^{b}	15.0^{b}
Chlorpyrifos	11.1^{b}	8.0^{c}	20.4	16.0	8.6	12.3	10.5^{ab}	14.3^{bc}	12.7^{c}
Absolute control	20.3^{ab}	20.2^{a}	20.4	20.1	20.2	20.2	20.9^{a}	22.2^{a}	20.5^{a}

*Means with the same letter in each column are statistically the same (Duncan $\alpha \leq 0.05$).
[1]Two days before the first application.
[2]Two days before the second application.
[3]Five days after the first application.

Table 4. Percentage of damage by leaf miner *Liriomyza sativae* Blanchard in leaf chickpea plants. Culiacan, Sinaloa, Mexico.

Treatments		Parasitoid species				Total
2013		*Neochrysocharis* spp.	*Opius* spp.	*Closterocerus* spp.	*Diglyphus* spp.	
	LLs					74
Chlorantraniliprole	Ptoid	0	7	2	1	10
	Ptism	0	9.45	2.70	1.35	14
	LLs					37
Cyromazine	Ptoid	0	1	1	1	3
	Ptism	0	2.70	2.70	2.70	8
	LLs					75
Spinosad	Ptoid	0	1	0	0	1
	Ptism	0	1.33	0	0	1
	LLs					184
Chlorpyrifos	Ptoid	0	6	0	1	7
	Ptism	0	3.26	0	0.54	4
	LLs					116
Absolute control	Ptoid	15	0	7	0	22
	Ptism	12.9	0	6.03	0	19
2014						
	LLs					325
Chlorantraniliprole	Ptoid	10	1	0	25	36
	Ptism	3.08	0.31	0	7.69	11
	LLs					242
Cyromazine	Ptoid	5	1	0	0	6
	Ptism	2.07	0.41	0	0	2
	LLs					318
Spinosad	Ptoid	2	2	0	2	6
	Ptism	0.63	0.63	0	0.63	2
	LLs					392
Chlorpyrifos	Ptoid	10	0	0	2	12
	Ptism	2.55	0	0	0.51	3
	LLs					240
Absolute control	Ptoid	10	0	0	20	30
	Ptism	4.1	0	0	8.3	12

LLs = Larvae of *L. sativae*, Ptoid = Parasitoid, and Ptism = Parasitism.

Table 5. Parasitoid species and parasitism (%) of larvae of *Liriomyza sativae* Blanchard in 400 chickpea leaves. Culiacan, Sinaloa, Mexico.

It was observed in 2013 that the number of adults of the leaf miner emerged from leafs decreased to 39, 39, 12, and 26%, with chlorantraniliprole, cyromazine, spinosad, and chlorpyrifos, with respect to 100% of the absolute control (**Table 3**). In 2014, it was decreased to 45, 66, 10, and 25%. The above was to be expected, since the same effect had been observed in the number of live larvae and empty mines. In this way, the results of this research can help increase local awareness to reduce the use of broad-spectrum insecticides [25].

The percentage of folioles damaged after application of insecticides in 2013 indicates that, with chlorantraniliprole, cyromazine, and spinosad, the damage decreased to 64, 63, and 59, respectively (**Table 4**), while chlorpyrifos was 63%. In 2014, the damages decreased to 44, 33, 27, and 38%, respectively, in relation to 100% of the absolute control.

The parasitoids obtained from the leaf miner of the chickpea were *Opius* spp. (Braconidae), *Diglyphus* spp., *Neochrysocharis* spp., and *Closterocerus* spp. (Eulophidae) (**Table 5**). In 2013, the leaf miner parasitism was 1% where spinosad was applied, 4% in plots treated with chlorpyrifos, 8% with cyromazine, and 14% in plots managed with chlorantraniliprole, compared to the average of 19% of observed parasitism in the control plot. In 2014, the parasitism was 2, 3, 2, and 11%, respectively, and in the control 12%. Three types of parasitoids found and the percentage of parasitism in the chickpea, with respect to what was observed in the absolute control, coincide with the results [26]; since 2006 and 2007, they found parasitoids *Opius monilicornis*, *Diglyphus crassinervis* and *Neochrysocharis ambitiosa*.

The estimate of net utility was determined by considering the value of production minus the cost of the crop, minus the value of the insecticides. The value of the ton of chickpea taken into account for operations was MX$ 12,700. In 2013, the highest production of chickpea was obtained where chlorantraniliprole was applied, with a net utility of MX$ 21,011, surpassing it to control with 71%, since its net utility was MX$ 12,305. With chlorpyrifos, spinosad, and cyromazine, a production was obtained that surpassed to control in 53, 48, and 37%, respectively.

In 2014, the highest production of chickpea was also obtained from the plots applied with chlorantraniliprole, from where a net utility of MX$ 16,036 was obtained, surpassing the control with 45%, whose net utility was MX$ 11,083. With chlorpyrifos, spinosad, and cyromazine, the respective increases were 39, 34, and 21%. Utility differences from 1 year to other may be due to the higher incidence and damage of *Liriomyza sativae* Blanchard in 2014.

4. Efficiency of diatomaceous earth for control of Mexican bean weevil (*Zabrotes subfasciatus* Boheman)

The results show that all doses of diatomaceous earth (DE) exerted an excellent control, such that in the evaluations registered at 15 days after application; 100% mortality was recorded in doses 4.0 and 5.0 g kg^{-1}, similar to those observed with the chemical control (deltamethrin). The doses of 2.0 and 3.0 g kg^{-1} of DE caused 95 and 96% mortality, in adults of *Zabrotes subfasciatus*. The dose of 1.0 g of DE caused 93% mortality, without statistical differences between the averages obtained with doses of DE and deltamethrin (**Table 6**), although these averages were significantly different from the average observed in the absolute control. The same behavior was observed in the evaluations recorded at 30, 45, and 60 daa (**Table 6**), where it can be seen that all DE doses used for pest control caused mortalities higher than 90%; this indicates that although the period of exposure was 2 months, mortality rates were maintained at 100% with the doses of 3.0, 4.0, and 5.0 g kg^{-1} of DE and deltamethrin at dose 1.0 mL kg^{-1} of seed. The lowest doses (1.0 and 2.0 g kg^{-1}) also exerted excellent control of *Zabrotes subfasciatus* Boh., with statistical differences in mortality only in relation to that obtained in the absolute control.

The results of the first experiment served to make the decision to perform a second experiment with lower doses that were 0.2, 0.4, 0.6, 0.8, and 1.0 g kg^{-1} DE, 0.1 mL kg^{-1} Deltamethrin (chemical) and an absolute control (without application). The results indicated that at 10 daa DE, the doses of 0.6, 0.8, and 1.0 g kg^{-1} of seed resulted in 100% mortality (**Table 7**), similar to that caused by the chemical control (deltamethrin), without significant differences between the averages. However, these mortality rates were significantly different from those at 0.2 and 0.4 g kg^{-1} of seeds, and even more with respect to the percentage of mortality (0) in the absolute control. At 20 daa, it was observed that where doses of 0.6, 0.8, and 1.0 g kg^{-1} of DE and 0.1 mL kg^{-1} of deltamethrin (chemical) were applied, mortality rates were 100% for adult weevil of the bean, but were not significantly different to the 95% that was achieved with the dose of 0.4 g kg^{-1} of DE. However, if they were statistically different from the mortality (28% less) of that was achieved with the dose of 0.2 g kg^{-1} of DE, likewise, with respect to the 0% observed in the absolute control.

At 40 daa, the treatments in doses of 0.6, 0.8, and 1.0 g kg^{-1} of DE and 0.1 mL kg^{-1} of deltamethrin, the mortality was 100% (**Table 7**), without significant difference with that caused by the dose of 0.4 g kg^{-1} of DE. However, these percentages were significantly different from the mortality that occurred with the dose of 0.2 g kg^{-1} of DE and with the absolute control.

Treatment/doses	**Mortality (%)**			
	15 dda	**30 dda**	**45 dda**	**60 dda**
Absolute control	1.2 b*	1.2 b	1.2 b	2.5 b
Deltamethrin/1.0 mL kg^{-1}	100 a	100 a	100 a	100 a
DE/1.0 g kg^{-1}	93.0 a	93.0 a	93.0 a	98.7 a
DE/2.0 g kg^{-1}	95.0 a	97.5 a	97.0 a	98.7 a
DE/3.0 g kg^{-1}	96.0 a	97.5 a	97.0 a	100 a
DE/4.0 g kg^{-1}	100 a	100 a	100 a	100 a
DE/5.0 g kg^{-1}	100 a	100 a	100 a	100 a

*Means with different letters in each column are statistically different, according to Tukey test ($\alpha \le 0.05$); dda = days after application.

Table 6. Percentage of adult mortality of bean weevil (*Zabrotes subfasciatus* Boh.) treated with diatomaceous earth (DE).

Treatment/doses	Mortality (%)			
	10 dda	20 dda	30 dda	40 dda
Absolute control	0.0 d*	0.0 c	0.0 c	0.0 c
Deltamethrin/0.1 mL kg^{-1}	100 a	100 a	100 a	100 a
DE/0.2 g kg^{-1}	62.0 c	72.0 b	79.0 b	83.0 b
DE/0.4 g kg^{-1}	90.0 b	95.0 a	97.0 a	98.0 a
DE/0.6 g kg^{-1}	100 a	100 a	100 a	100 a
DE/0.8 g kg^{-1}	100 a	100 a	100 a	100 a
DE/1.0 g kg^{-1}	100 a	100 a	100 a	100 a

*Means with different letters in each column are statistically different, according to Tukey test ($\alpha \leq 0.05$); dda = days after application.

Table 7. Percentage of adult mortality of bean beetle (*Zabrotes subfasciatus* Boh.) treated with lower doses of diatomaceous earth (DE).

Seed germination was similar with all treatments applied, including the absolute control, with a seedling emergence ranging from 94 to 96%, considered as normal, and it was assumed that the diatomaceous earth had no effect on the seed germination.

These results coincide with the results of Mikami et al. [27], where it is pointed out that diatomaceous earth is a mineral with insecticidal potential against the bean weevil, applied in doses of 1.0 g kg^{-1} to have a 100% mortality of the 3–8 days after application. They also coincide with those of [12, 14, 28, 29], since they report that these inert powders have been used with great success in controlling large numbers of stored grain pests, among which are *Oryzaephilus surinamensis*, *R. dominica*, *Tribolium castaneum*, *T. confusum*, *Cryptolestes ferrugineus*, *S. zeamais*, *S. granarius*, *S. oryzae*, *Prostephanus truncatus*, *Acanthoscelides obtectus*, and *Zabrotes subfasciatus*. It is reported that diatomaceous earth doses of 0.5, 1.0, and 2.0 kg per ton of maize seed alone and combined with deltamethrin synergized with piperonyl butoxide cause mortality higher than 97% of maize weevil, up to 120 days after application [30], and this same behavior was observed when it was combined with the insecticide deltamethrin.

In addition, they agree with those of [31], because they indicate that diatomaceous earth is an alternative for the control of *Zabrotes subfasciatus* Boheman, since after 5 days of exposure and temperatures of 27–30°C, they had mortality of 100% with all the applied doses (0.5, 0.75, and 1.0 g kg^{-1} of seed), concluding that the suitable doses for the control of this pest of the store are those of 0.75 and 1.0 g kg^{-1} of seed.

Mineral powders such as zeolite can control stored grain pests such as *Sitophilus oryzae*, *Tribolium confusum*, and *Oryzaephilus surinamensis*, and that therefore, this material can be successfully used as a grain and seed protector [32]. Likewise it is reported that inert dusts cause abrasive effects on the cuticle of insects, resulting in loss of water and, consequently, death [33, 34]. In addition, these powders may be used in combination with other products, such as vegetable powders to increase the efficacy of pest control.

5. Conclusions

The use of biorational insecticides is a good alternative for the control of *Liriomyza sativae* Blanchard in chickpea. While that in bean, the doses of 0.8–5.0 g kg^{-1} of diatomaceous earth efficiently controlled the Mexican bean weevil, but the recommended dose is 0.8 g kg^{-1} of seed, since with this dose, it can be controlled with sustainability and does not affect seed germination, as with the other doses evaluated.

Author details

Jacobo Enrique Cruz Ortega, Leopoldo Partida Ruvalcaba*, Raymundo Medina López, Tomás Díaz Valdés, Teresa de Jesús Velázquez Alcaraz and Felipe Ayala Tafoya

*Address all correspondence to: arpolo@yahoo.com.mx

Faculty of Agronomy, Autonomous University of Sinaloa, Culiacan Rosales, Sinaloa, Mexico

References

[1] Acharjee S, Sarmah BK. Biotechnologically generating super chickpea for food and nutritional security. Plant Science. 2013;**207**:108-116. DOI: 10.1016/j.plantsci.2013.02.003

[2] FAO. FAOSTAT, Agriculture. Food and Agriculture Organization, FAO Statistical Year Book [Internet]. 2013. Available from: http://www.faostat3.fao.org/browse/Q/*/E [Accessed: YYYY-MM-DD]

[3] Salinas PRA, Cortez ME, Macías CJ. Guía para producir garbanzo en el norte de Sinaloa. INIFAP-CIRNO. Campo Experimental Valle del Fuerte [Internet]. Los Mochis, Sinaloa, México: Folleto Técnico 29; 2008. Available from: http://biblioteca.inifap.gob.mx:8080/xmlui/bitstream/handle/123456789/1662/Guia%20para%20producir%20garbanzo%20en%20el%20norte%20de%20Sinaloa.pdf?sequence=1 Accessed: YYYY-MM-DD

[4] Peterson C, Tsao R, Eggler AL, Coats JR. Insecticidal activity of cyanohydrin and monoterpenoid compounds. Molecules. 2000;**5**:648-654 http://dx.doi.org/10.3390/50400648

[5] Oliva A, Kimudini M, Wedge ME, Harries DD, Hale L, Aliotta AG, Duke MO. Natural fungicides from *Ruta graveolens* L. leaves, including a new quinolone alkaloid. Journal of Agricultural and Food Chemistry. 2003;**51**:890-896. DOI: 10.1021/jf0259361

[6] Miyazawa M, Nakamura Y, Ishikawa Y. Insecticidal sesquiterpene from *Alpinia oxyphylla* against *Drosophila melanogaster*. Journal of Agricultural and Food Chemistry. 2000;**48**:3639-3641. DOI: https://doi.org/10.1021/jf000325z

[7] Morimoto M, Tanimoto K, Nakano S, Ozaki T, Nakano A, Komai K. Insect antifeedant activity of flavones and chromones against *Spodoptera litura*. Journal of Agricultural and Food Chemistry. 2003;**51**:389-393. DOI: 10.1021/jf025627a

[8] González J, Reyes F, Salas C, Santiago M, Codriansky Y, Coliheuque N, Silva H. *Arabidopsis taliana*: A model host plant to study plant pathogen interaction using Chilean field isolates of *Botrytys cinerea*. Biological Research. 2006;**39**:221-228 http://dx.doi.org/10.4067/S0716-97602006000200004

[9] SIAP. Servicio de Información Agroalimentaria y Pesquera [Internet]. 2016. Available from: www.gob.mx/siap/ [Accessed: YYYY-MM-DD]

[10] CIAT. Insectos del frijol almacenado y su control por CIAT (Centro Internacional de Agricultura Tropical). Cali, Colombia: Folleto de divulgación 1344; 1986. p. 4

[11] Soto POJ. Evaluación de la resistencia de variedades de frijol común (*Phaseolus vulgaris* L.) al ataque de gorgojo del frijol *Zabrotes subfasciatus* Boheman [Internet thesis]. Escuela Agrícola Panamericana: Zamorano, Honduras; 2014. Available from: https://bdigital.zamorano.edu/bitstream/11036/3377/1/AGI-2014-T041.pdf Accessed: YYYY-MM-DD

[12] Arthur FH. Toxicity of diatomaceous earth to red flour beetles and confused flour beetles (Coleoptera: Tenebrionidae): Effects of temperature and relative humidity. Journal of Economic Entomology. 2000;**93**:526-532. DOI: 10.1603/0022-0493-93.2.526

[13] Arthur FH. Immediate and delayed mortality of *Oryzaephilus surinamensis* (L.) exposed on wheat treated with diatomaceous earth: Effects of temperature, relative humidity, and exposure interval. Journal of Stored Products Research. 2001;**37**:13-21. DOI: 10.1016/S0022-474X(99)00058-2

[14] Korunic Z. Rapid assessment of the insecticidal value of diatomaceous earths without conducting bioassays. Journal of Stored Products Research. 1997;**33**:219-229. DOI: 10.1016/S0022-474X(97)00004-0

[15] Korunic Z. Diatomaceous earths: A group of natural insecticides. Journal of Stored Products Research. 1998;**34**:87-97 https://doi.org/10.1016/S0022-474X(97)00039-8

[16] INEGI. Anuario estadístico de Sinaloa. Instituto Nacional de Estadística y Geografía [Internet]2010. Available from: www.inegi.org.mx/est/contenidos/espanol/sistemas/aee10/info/sin/mapas.pdf Accessed: YYYY-MM-DD

[17] Avilés GM, Fú AA, Pérez VCJ, Wong PJJ, Garzón TJA. Los insectos plaga del garbanzo en el estado de Sinaloa. Descripción y Manejo. INIFAP-CIRNO-CEVACU. Culiacán, Sinaloa, México: Folleto Técnico 27; 2004

[18] Palacios TRE, Romero NJ, Étienne J, Carrillo JL, Valdez JM, Bravo CH, Koch SD, López MV, Terán AP. Identificación, distribución y plantas hospederas de diez especies de Agromyzidae (Insecta: Díptera) de interés agronómico en México. Acta Zoológica Mexicana. 2008;**24**:7-32

[19] Spencer KA, Stegmaier CE. Agromyzidae Of Florida, With A Supplement On Species From The Caribbean. Arthropods Of Florida And Neighboring Land Areas [Internet]. 1973. Available from: http://www.freshfromflorida.com/Divisions-Offices/Plant-Industry/Florida-State-Collection-of-Arthropods/FSCA-Publication-Archive/Arthropods-of-Florida-and-Neighboring-Land-Areas [Accessed: YYYY-MM-DD]

[20] Spencer KA, Steyskal JC. Manual of the Agromyzidae (Diptera) of the United States Agriculture Handbook [Internet]. 1986. Available from: http://naldc.nal.usda.gov/naldc/download.xhtml?id=CAT86871406&content=PDF [Accessed: YYYY-MM-DD]

[21] Wharton RA. Subfamily Opiinae. In: Wharton RA, Marsh PM, Sharkey MJ, editors. Manual of the New World Genera of the Family Braconidae (Hymenoptera). Special Publications of the International Society of Hymenopterists 1. 1997

[22] La Salle J, Parella MP. The chalcidoid parasites (hymenoptera, Chalcidoidea) of economically important *Liriomyza* species (Diptera, Agromyzidae) in North America. Proceedings of the Entomological Society of Washington. 1991;**93**:571-591

[23] SAS Institute. SAS/STAT® 9.1 User's Guide. Cary, NC: SAS Institute, Inc.; 2004. 1028 p

[24] Abbott WS. A method of computing the effectiveness of an insecticide. Journal of Economic Entomology. 1925;**18**:265-267. DOI: 10.1093/jee/18.2.265a

[25] Vázquez-Luna A, Pérez-Flores L, Díaz-Sobac R. Biomoléculas con actividad insecticida: una alternativa para mejorar la seguridad alimentaria. Ciencia y Tecnologia Alimentaria. 2007;**5**:306-313

[26] Cikman E, Kaplan M. Effects of azadirachtin a [*Azadirachta indica* a Juss (Meliacea)] on larval serpentine leafminers *Liriomyza cicerina* (Rondani, 1875) (Díptera: Agromyzidae) in chickpea. Journal of Applied Science Research. 2008;**4**:1143-1148

[27] Mikami AY, Pissinati A, Fagotti D, Menezes Junior AO, Ventura MU. Control of the Mexican bean weevil *Zabrotes subfasciatus* with kaolin. Ciência Rural. 2010;**40**:1497-1501. DOI: 10.1590/S0103-84782010005000108

[28] Cook DA, Armitage DM. Efficacy of a diatomaceous earth against mite and insect populations in small bins of wheat under conditions of low temperature and high humidity. Pest Management Science. 2000;**56**:591-596. DOI: 10.1002/1526-4998(200007)56:7<591::AID-PS180>3.0.CO;2-L

[29] Fields P, Korunic Z. The effect of grain moisture content and temperature on the efficacy of diatomaceous earths from different geographical locations against stored-product beetles. Journal of Stored Products Research. 2000;**36**:1-13. DOI: 10.1016/S0022-474X(99)00021-1

[30] Mazzuferi VE, Goncalvez RH, Tablada M, García D. Efectividad y persistencia de la tierra de diatomeas en el control de *Sitophilus zeamais* (Coleoptera: Curculionidae) en semillas de maíz y su incidencia en la calidad. Boletín Sanidad Vegetal. Plagas. 2006;**32**:363-371

[31] Lazzari FN, Ribeiro CCS. Control of *Zabrotes subfasciatus* (Boheman) (Coleoptera, Chrysomelidae, Bruchinae) in *Phaseolus vulgaris* Linnaeus, using diatomaceous earth under different temperatures. In: Proceedings of the 9th International Working Conference on Stored Product Protection. Campinas, Brazil; 15–18–10-2006. p. 804-810

[32] Rumbos CI, Sakkab M, Berillisc PC, Athanassiou G. Insecticidal potential of zeolite formulations against three stored-grain insects, particle size effect, adherence to kernels and

influence on test weight of grain. Journal of Stored Products Research. 2016;**68**:93-101. DOI: 10.1016/j.jspr.2016.05.003

[33] Subramanyam B, Roesli R. Inert Dust. In: Subramanyamy BD, Hagstrum W, editors. Alternatives to Pesticides in Stored Product IMP. Kluwer Academic Publishers; 2000. p. 321-380. DOI: 10.1007/978-1-4615-4353-4_12

[34] Silva AG, González GP, Hepp GR, Casals BP. Control de *Sitophilus zeamais* Motschulsky con polvos inertes. Agrociencia. 2004;**38**:529-536

Chapter 2

Plant Lectins with Insecticidal and Insectistatic Activities

Edgar Antonio Reyes-Montaño and
Nohora Angélica Vega-Castro

Additional information is available at the end of the chapter

http://dx.doi.org/10.5772/intechopen.74962

Abstract

Lectins are an important group of proteins which are spread in all kingdoms of life. Their most lighted characteristic is associated to their specific carbohydrate binding, although function has been not even identified. According to their carbohydrate specificity, several biological activities have been assessed, finding that lectins can be used as mitogenic agents, biomarkers, and cytotoxic and insecticide proteins. Lectins have been classified according to several features such as structure, source, and carbohydrate recognition. The Protein Research Group (PRG) has worked on Colombian seeds from the family of Fabaceae and Lamiaceae plants, isolating and characterizing their lectins, and found more than one lectin in some plants, indicating that according to its specificity, different lectins can have different biological activities. In the case of legume domain lectins, they have shown the biggest potential as insecticide or insectistatic agents due to the glycosylation pattern in insect midgut cells. This review attempts to identify the characteristics of plant legume lectin domains that determine their insecticidal and insectistatic activities.

Keywords: lectin, insecticide, insectistatic, legume

1. Introduction

Lectins are glycoproteins of nonimmune origin that recognize and bind carbohydrates. These proteins are found in a wide variety of species (viruses, bacteria, fungi, seaweed, animals, and plants). This review is mainly based on information of plant lectins that have been found as important new agents in biological control. Plant lectins have been widely studied, and in this group, the legume lectins have been related to insecticidal and insectistatic activities. In addition, *Phaseolus vulgaris* (PHA), *Glechoma hederacea* (Gleheda), *Canavalia ensiformis* (ConA), *Griffonia simplicifolia* (GSII), and *Pisum sativum* (PSA) lectins and other legume and Lamiaceae lectins have been studied by the Protein Research Group (PRG) in Colombia. It was evidenced that plant legume lectin domains have structural features characterized by a high percentage of β-sheet structures associated with dimeric or tetrameric assembly, presenting several specific sugar recognition sites, including mannose. In addition to these features, these lectins can interact with the digestive system of insect pests and produce a decrease in intestinal absorption capacity.

2. Definition, classification, and general features of lectins

Lectins are proteins or glycoproteins of the nonimmune origin with specific binding affinity for the carbohydrate moiety of glycoconjugates [1]. Lectins comprise a structurally diverse class of proteins characterized by their ability to selectively bind carbohydrate moieties of the glycoproteins of the cell surface. Lectins may be obtained from plant, microbial, or animal sources and may be soluble or membrane bound [2]. In nature, lectins play a role in biological recognition phenomena involving cells and proteins and thereby protect plants against external pathogens such as fungi and other organisms. The ability to bind and agglutinate red blood cells is well known and used for blood typing; hence, the lectins are commonly called hemagglutinins [3].

The term lectin is derived from the Latin word *legere* meaning "to choose" or "select" and has been generalized to encompass all nonimmune carbohydrate-specific agglutinins regardless of blood type specificity or source. Lectins were initially found and described in plants, but in subsequent years, multiple lectins were isolated from microorganisms and also from animals [4]. Interestingly, plant and animal lectins show no primary structural homology, but they demonstrate similar preferential binding to carbohydrates [5]. This suggests that animal and plant lectin genes may have coevolved, thus highlighting the importance of lectin-carbohydrate interactions in living systems [6].

Based on the amino acid sequences of available lectins, it is deduced that the carbohydrate-binding property of most lectins resides in a polypeptide sequence, which is termed as "carbohydrate-recognition domain" [7]. The binding with simple or complex carbohydrate conjugates is reversible and non-covalent. The specificity of lectins toward carbohydrates can be defined on the basis of "hapten inhibition test," in which various sugars or saccharides are tested for their capacity to inhibit the property of hemagglutination of erythrocytes [8].

Lectins have been classified according to different features such as source (animal, vegetal, fungal, viral), carbohydrate affinity (mannose, glucose, galactose, fucose, sialic acid), number, and specificity of carbohydrate recognition domains (merolectins, hololectins, chimerolectins, and superlectins) [9]. However, current classification is based on 3D structure and is related to 48 families (**Table 1**) [10].

http://dx.doi.org/10.5772/intechopen.74962

Distribution								
No.	Family	Fold	Assembly	Animal	Plant	Fungi	Bacteria	Virus
1	L-type	Jelly roll	Dimer	x	x	x		
	L-type-like	Jelly roll	Monomer	x	x	x		
2	Galectin	Jelly roll	Monomer, dimer	x		x		x
3	Pentaxim	Jelly roll	Pentamer	x				
4	I-type	Ig-like β-sandwich	Linked to different domains	x				
5	C-type	α/β-fold	Linked to different domains	x				
6	Hyaladherin	α/β-fold	Linked to different domains	x				
7	Chitinase-like	$(\beta/\alpha)_8$-Barrel	Monomer	x	x	x	x	x
8	M-type	$(\alpha/\alpha)_7$-Barrel	Monomer	x	x	x	x	
9	R-type	β-Trefoil	Linked to enzyme	x	x	x	x	
	R-type-like	β-Trefoil	Linked to different domains			x	x	
10	ACA-like	β-Trefoil	Dimer		x			
11	Botulinum neurotoxin-like	β-Trefoil	Linked to different domains				x	
12	F-box	Jelly roll	Linked to different domains	x				
13	F-type	Jelly roll	Linked to different domains	x	x	x	x	
14	PA-LL-like	Jelly roll	Dimer				x	
15	P-type	α/β-fold	Dimer	x				
16	Ficolins	Fibrinogen-like	Trimer	x				
17	Malectin	Jelly roll	Monomer	x				
18	Calnexin	Jelly roll	Monomer	x				
19	Tachylectin-2-like	5-Bladed β-propeller	Monomer	x				
20	Tachycitin-like	β-sheet-cysteine fold	Monomer	x				
21	Hevein	Cystine-knot motif	Dimer	x	x			
22	Jacalin-related	β-Prism I	Tetramer	x	x			
23	SUEL-related	α/β-fold	Linked to different domains	x				

Distribution								
No.	Family	Fold	Assembly	Animal	Plant	Fungi	Bacteria	Virus
24	H-type	Six-stranded antiparallel β-sandwich	Hexamer	x	x			
25	Cystine-knot	Cystine-knot motif		X				
26	TgMIC4	α/β-fold	Tandem repeat	x				
27	TgMIC1	Sialic acid binding protein	Linked to different domains	x				
28	LysM	βααβ-Motif	Triple repeat	x	x	x	x	
29	LNP-type	α/β-fold	Monomer	x	x		x	
30	Monocot	β-Prism II	Monomer, dimer, tetramer		x		x	
31	ABL-like	α/β-sandwich	Dimer, tetramer		x	x		
32	CV-N	Three-stranded β-sheet and β-hairpins	Monomer		x	x	x	
33	PVL-like	Seven-bladed β-propeller	Monomer			x		
34	AAL-like	Six-bladed β-propeller	Monomer			x	x	
35	Flocculins	β-Sandwich	Monomer			x	x	
36	PCL-like	Jelly roll	Tandem repeat			x		
37	BC2LCN	Jellyroll	Trimer				x	
38	Staphylococcal toxin	β-Barrel	Monomer				x	
39	AB5 toxin	α/β-fold	AB5				x	
40	PA-IIL-like	β-Sandwich	Dimer				x	
41	MVL	α/β-fold	Dimer				x	
42	PapG	β-Sandwich	Linked to different domains				x	
43	FimH	β-Sandwich	Linked to different domains				x	
44	F17-G	β-Sandwich	Linked to different domains				x	
45	Hemagglutinin	Jelly roll	Trimer					x
46	RotavirusVP4	Jelly roll	Virus capsid					x
47	Viral proteins	β-Sandwich	Virus capsid					x
48	Knob domain	Jelly roll	Virus capsid					x

Folding, assembly, and source of each family is shown.

Table 1. Lectin families in nature.

3. Structure and biological activities of plant lectins

Lectins are mainly present in seeds of plants [4, 8, 9], but they are also identified in vegetative tissues such as bulbs, tubers, rhizomes, roots, bark, stems, fruits, and leaves [11].

As previously mentioned, based on their number domains and their characteristics, plant lectins can be divided into four classes [9]:

- **Merolectins** are lectins that possess a single carbohydrate-binding domain. As a result, the merolectins do not present agglutinating activity.
- **Hololectins** contain two or multivalent carbohydrate-binding sites.
- **Chimerolectins** possess a carbohydrate-binding domain and an additional domain that confers other biological activities.
- **Superlectins** are lectins with two or multivalent carbohydrate domains that are able to recognize structurally unrelated sugars.

However, since 1998, five novel lectin domains have been identified in plants. At present, plant lectins are classified into 12 different families, with distinct carbohydrate-binding domains. The families are *Agaricus bisporus* agglutinin homologs, amaranthines, class V chitinase homologs, *Euonymus europaeus* agglutinin family, *Galanthus nivalis* agglutinin family, proteins with hevein domains, jacalins, proteins with legume lectin domains, LysM domain proteins, the *Nicotiana tabacum* agglutinin family, and the ricin B family [12].

In general, the three-dimensional structure of lectins is composed of a high content of β-sheets with little contribution from α-helixes. The β-sheets are connected by loops forming antiparallel chains. The stability of dimers and tetramers is conferred by hydrophobic interactions, hydrogen bonds, and salt links [13]. Three regions are formed in carbohydrate-binding site [12–14]:

- The central region is constituted by a conserved core in which residues interact with metallic ions (Mg^{2+}, Mn^{2+}, and Ca^{2+}), required for carbohydrate interactions. This core provides necessary binding energy, but it is not important to the lectin's carbohydrate specificity.
- Some aromatic residues surround the core and occupy variable positions in a horseshoe shape. This region is fully involved in the lectin's monosaccharide specificity.
- Finally, residues with higher variability are located in the outer zone and are involved in interactions with larger oligosaccharide ligands.

The structural features of plant lectins are shown in **Figure 1**, which is possible to see the high content of β-sheets (**Figure 1A**) and the structure of a typical carbohydrate recognition domain (**Figure 1B**).

However, the kind of expressed lectins can have some differences according to the specific tissue or the moment in which the plant is expressing it. A lot of plant lectins are constitutively expressed in high amounts in seeds and vegetative storage tissues where

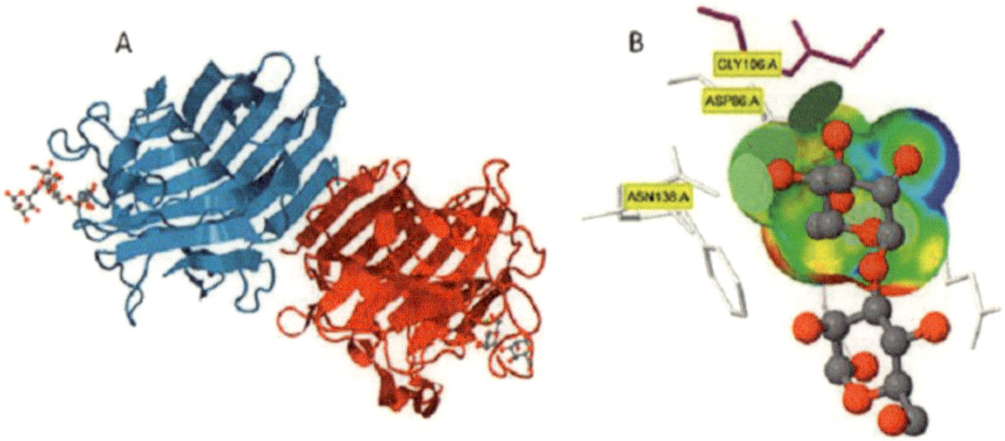

Figure 1. Structural conformation of plant lectins. (A) *Pterocarpus angolensis* homodimer lectin (PDB code (2PHF)). The β-sheet conformation is the most usual in plant lectins (β-sandwich). (B) The carbohydrate recognition domain (CRD) is highly conserved in plant lectins, according to its specificity.

they have been shown to play a role in plant defense [15]. But, plants also express minute amounts of specific lectins as particular responses toward environmental stresses and pathogen attack. In the absence of plant stress, the inducible lectins are not expressed at detectable levels [16]. According that, a central question which has often been asked but up to now not yet been answered definitively is that on the biological function(s) of plant lectins. Several functions have been mentioned, but there is not a final decision about that. However, because of its carbohydrate interactions, lectins have been tested for several biological functions, getting interesting results in some of them. Biological activities are related to immunomodulatory and antitumor [17–19], antifungal [20–23], antiparasitic [24–26], antiproliferative [27–30], healing process [31–33], drug delivery [34–36], as histochemical markers [37–39], biosensors [40, 41], insecticide [42–46], etc.

4. Fabaceae (legume) and Lamiaceae (mint) lectins

The specific carbohydrate recognition shown by lectins makes them important tools in glycobiology, and, although their physiological role remains unknown, they appear to mediate protein-cell and cell-cell interactions. Lectins are widespread in nature, and most of them have been isolated and characterized from Fabaceae, Gramineae, and Lamiaceae families, among others [47, 48]. Those lectins have been related to insect defense mechanisms, storage proteins, carbohydrate transport, mechanisms of physiological regulation, and mitogenic stimulation processes [49–55]. The ability of the nitrogen-fixing bacteria rhizobia to form a symbiotic relationship with legumes, in which plant root lectins are involved, is well known. The plant-associated bacteria have important effects on plant health and productivity [56–59]. Thus biofilm formation on plants is associated with symbiotic and pathogenic responses, and some root lectins promote this process [60]. The lectins could be a good biotechnological alternative in the control of bacterial biofilms for different purposes, for example, clinical applications [61]. In general, plant lectins have been widely used for studying carbohydrates on cell surface, for typing blood groups, isolating glycoconjugates, and detecting changes in normal oligosaccharide synthesis in tumoral disorders and other pathologies [62–66].

Lectins from Fabaceae have been extensively studied and have a broad specificity for any carbohydrate moieties regardless of having highly conserved amino acid sequences between different species. These proteins have been for a long time a paradigm in the research of interaction protein-carbohydrate and their relationship structure-function [67, 68]. Available sequences (RCSB PDB, UniProtKB/Swiss-Prot) show 20% similarity and 20% of identical amino acids, and conserved amino acids are in the "binding site" and coordinate metal ions [9]. These proteins generally have two or four identical subunits with a molecular weight around 25 kDa; each one contains a binding site for metal ions. A typical example of dimeric lectins belongs to the Viceae tribe. The tetrameric lectins are present in species of the tribe Diocleae, specific by glucose/mannose. In these tribes, many lectins have been isolated and characterized with some biochemical differences and molecular similarities [47]. Recently, subtribe Diocleinae in the Millettioid legumes have been taxonomically tangled together with the large heterogeneous tribe Phaseoleae; however, a comprehensive molecular phylogenetic analysis based on nuclear and chloroplast markers includes all genera ever referred to Diocleae except for the monospecific Philippine *Luzonia*, resolving several key generic relationships within the Millettioid legumes and considered classification of Diocleinae subtribe as a tribe with three main clades: *Canavalia, Dioclea*, and *Galactia*. *Canavalia* clade has species gender *Canavalia*; *Dioclea* clade includes *Dioclea, Cymbosema, Cleobulia* and *Macropsychanthus*; and *Galactia* clade gender has *Galactia, Neorudolphia, Rhodopsis, Bionia, Cratylia, Lackeya, Camptosema*, and *Collaea* [69].

This tribe is widely distributed throughout the neotropics, and several species from the genus *Dioclea* have been shown to possess a lectin closely related to ConA (lectin type I). The better characterized lectins have been those from *D. grandiflora* [70, 71], *D. lehmanni* Diels [72], and *D. sericea* Kunth [73], among others, all of them belong to the Man/Glc group; their physicochemical properties and structural features are very similar [74].

Studies carried out in the PRG have allowed us to find other lectins having distinct structural and functional properties (named lectin type II) from *Diocleae lehmanni* (DLL), *Dioclea sericea* (DSL), *Dioclea grandiflora* (DGL), *Canavalia ensiformis* (CEL), and *Galactia lindenii* (GLL) [73, 75–77]. These lectins are localized in the same cellular compartment as happens in *D. lehmanni* seeds [78] and have different physicochemical properties; this allow us to question about the physiological role of these proteins. Lectin type II has high affinity toward H type 2 blood group (α-L-Fuc (1–2)-β-D-Gal (1–4)-β-D-GlcNAc-O-R), and the N-terminal region presents a unique sequence hitherto found in some Diocleinae lectins and suggests a functional similarity among this type of lectin which possesses distinctive characteristics differentiating them from "classical" mannose/glucose (Man/Glc) lectins. Taking subunit MW into account, it has been demonstrated that tetrameric forms prevailed in type I lectins, being in fast equilibrium with dimers and monomers whose amount depended upon pH or solution ionic strength [79], while some lectins from type II prevalence dimeric forms (**Table 2**). Despite their high similarity, these ConA-like (type II) lectins could induce different responses in biological assays; for example, when tested for stimulation of human lymphocyte proliferation in vitro, ConBr had a higher proliferation index than ConA, possibly due to minor changes in binding specificities [80].

Type	Species	Specificity	Monosaccharide inhibitor	Erythroagglutination	Native (kDa)	Subunits (kDa)	pI	References
I	*D. grandiflora*	Man/Glc	Man, Glc, Fru	Rabbit	100	α:25–α:26; β:13–β:14; γ:8–γ:9	8.6–9	[70, 71]
	D. lehmanni		Man, Glc, Fru, L-sorbose, Me-α-D-Man, Me-α-D-Glc, trehalose	Rabbit, A+, O+, B+		α:25.3; β:14; γ:N.D	8.0–8.4	[72]
	D. sericea		Man, Glc	A+, O+, B+	57.7	α:29.9; β:16.5; γ: 13.4	6.6–6.9	[73]
	D. altisima		Man, Glc, Fru	Rabbit	100	α:26.3; β:14; γ: 9	8.6–9.0	[131]
	D. violaceae		Man, Glc, Fru, maltose	Rabbit		α:29.5; β:15.8; γ: 11.7		[132]
	D. rostrata		Man, Glc, Fru	Rabbit, O+ and B+		α:30.9; β:15.8; γ: 11.7		[67]
	D. lasiophylla		Man, Me-α-D-Man, ovalbumin, fetuin	Rabbit		α:25,569; β:12,998; γ: 12,588		[133]
	D. sclerocarpa		Glc; Gal	Rabbit	102	α: 25,606; β:12,832; γ:12,752		[134]
	C. ensiformis		Man, Me-α-fructofuranoside	Rabbit	96	α:25.5; β:14; γ:12.5	7.1	[67]
	C. mollis		Glc, Me-α-D-Man	Rabbit > A+, O+, B+		α:30; β:16; γ: 14	8.5–8.6	[135]
	C. roseum		Man	Rabbit		α:30; β:18; γ: 12		[136]
	G. lindenii		p-Nitrophenyl-β-D-mannopyranoside, Man	A+, O+	100	29; 60	6,5	[77]

http://dx.doi.org/10.5772/intechopen.74962

Type	Species	Specificity	Monosaccharide inhibitor	Erythroagglutination	Native (kDa)	Subunits (kDa)	pI	References
II	*C. ensiformis*	H-Type II	Sucrose, melezitose, lactose	A+, O+, B+	57.5	29–30	5.2–5.4	[76]
	D. grandiflora		Sucrose, melezitose, lactose	A+, O+, B+	58.9	29–30	5.1–5.4	[76]
	D. lehmanni		Sucrose, melezitose, lactose	A+, O+, B+ > rabbit	58.4	29–30	6.5–6.6	[75]
	D. sericea		Lactose, sucrose, melibiose	A+, O+, B+	57.27	26.58–30	5.3–5.7	[73]
	G. lindenii		GalNAc, Me-β-Gal, Lactose	B+, O+ > A+	104,256	26,064	8.3	[137]
	C. roseum		GalNAc and N-acetyl-α-D-lactosamine	Rabbit	65	29	—	[138]
	Captosemin		N-acetyl-α-D-galactosamine	A+, O+, B+	104	26	—	[139]

Abbreviations: kDa, kilodalton; pI, isoelectric point; H-type II, antigen (α-L-Fuc(1–2)-β-D-Gal(1–4)-β-D-GlcNAc-O-R); Man, mannose; Glc, glucose; Me, methyl; Gal, galactose; Fru, fructose; GalNAc, N-acetyl-α-D-galactosamine.

Table 2. Physicochemical properties of lectins of Diocleae tribe.

Lamiaceae lectins have been little studied despite preliminary reports on their ability to recognize the Tn/T antigens [81], normally a cryptic structure in the peptide core of O-glycoproteins and which is widely expressed in several tumors and other disorders such as Tn syndrome and IgA nephropathy [82–85]. The importance of Thomsen-Friedenreich antigen (TF or T, galactose (Gal) β1,3 GalNAcα-O-serine (Ser)/threonine (Thr)) as well as to its precursor, the Tn antigen, and its sialylated forms (sTn) has been reviewed recently [86–91]; according to the above, it is important to have alternatives to study these structures such as the lectins and antibodies. However, a word of caution should be given as accumulating evidence, which has shown that mAbs and lectins do not interact with Tn-containing structures in an identical manner. The observed differences have been ascribed to different Tn-density requirements for the interaction to occur [92].

Detailed studies have been carried out on a very few Lamiaceae species from the Northern hemisphere's temperate zone until now [93–97], and the lectin from *Salvia sclarea* L. seeds (SSL) was the first to be isolated and partially characterized [94]. By contrast, species from the Neotropical *Salvia* subgenus Calosphace Benth have been little explored despite their great diversity. A systematic survey has been conducted on species belonging to the Neotropical Calosphace Benth subgenus [98], and certain species naturalized in the New World have also been investigated [99], some having commercial value. Given the abundance of Lamiaceae species in Colombia and the potential biotechnological applications, our group undertook a systematic search for the identification, isolation, and characterization of lectins from selected species with the determination of their biological activities. The lectins from *S. palifolia* Kunth and *Hyptis mutabilis* (Rich.) Briq. [100] have been partially characterized, and a detailed work has been done with *S. bogotensis* Benth and *Lepechinia bullata* (Kunth) Epling [101, 102].

The importance of these proteins as tools in a variety of biological studies and detection, isolation, structural, and functional properties has been studied, and more recently, T/Tn-specific lectins have been found in the families Amaranthaceae, Fabaceae, Moraceae, and Orchidaceae, among others. The lectins themselves belong to five families of structurally and evolutionarily related proteins (amaranthines, legume lectins, jacalin-related lectins, type 2 ribosome-inactivating proteins, and GNA-related lectins) [103].

Interestingly, a lectin type I was found in *S. bogotensis* Benth. (SBoL-I) and *Lepechinia bullata* (Kunth) Epling (LBL-I) (such as those found in the tribe Diocleae type I), which recognizes mannose/glucose residues; this fact, together with the molecular properties and highly similar N-terminal regions, led us to propose that lectins type I and type II are two good differentiated groups with structural features proper of legume lectins family, particularly from Diocleae tribe, *Salvia,* and *Lepechinia* genders (**Table 3**) [104]. For these lectins, SDS-PAGE profile was similar to other mannose lectins, a band around 30 kDa with an isoelectric point near to 6.5, and they were able to agglutinate human RBCs from A, B, and O donors. This means that specificity by mannose/glucose moieties or mannose-rich glycan is not a unique feature of any family; conversely, species such as *Galanthus nivalis* (tribe Galantheae) [105] and *Centrolobium microchaete* (tribe Dalbergieae) [106], among others, even species from other families such as Moraceae have mannose/glucose lectins [107].

http://dx.doi.org/10.5772/intechopen.74962

Molecular properties	GLL-I[1]	DLL-I[2]	CRL-I[3]	CEL-I[4]	SBoL-I[5]	LBL-I[6]
M_r subunit (kDa)[7]	29	25, 14	ND	26.5	30–33	30–34
M_r protein (kDa)[8]	100	ND	ND	106	ND	ND
SDS-page (kDa)	29, 60	25, 14	30, 18, 12	26, 14, 12.5	30, 60	30, 60
Glycosylation	Si	ND	ND	No	Si	Si
Neutral Sugars (%)	ND	1.7–1.9	ND	ND	ND	ND
Isoelectric point (PI)	6.15	8.0; 8.13 8.3; 8.42	ND	7.1	6.5	6.5
Mannose inhibition (mM)	150	50	19.5	ND	ND	ND
Sequence N-terminal	ND	ADTIVAVELD SYPNTDIGDPSYPH	ADTIVAVELD SYPNTDIGDPSYPH	ADTIVAVELD TYPNTDIGDPSYPH	ADTIVAVELD	ADTIVAVELD

[1] *Galactia lindenii* lectin type -I (GLL-I) [77].
[2] *Dioclea lehmanni* lectin type I (DLL-I) [72].
[3] *Cymbosema roseum* lectin type I (CRL-I) [136].
[4] *Canavalia ensiformis* concanavalin A (CEL-I) [67].
[5] *Salvia bogotensis* lectin type I (SBoL-I) [104].
[6] *Lepechinia bullata* lectin type I (LBL-I) [104].
[7] Reduced conditions.
[8] Non-reduced conditions without heat.
ND, non-determined.

Table 3. Molecular properties of lectins type I from Fabaceae and Lamiaceae families.

5. Insecticide and insectistatic activity of plant lectins

There are several evidences for the defensive role of vegetal lectins in protecting plants against insect pests [108–110], and lectins are currently receiving a significant interest as insecticidal agents against sap-sucking insects including aphids and leaf and plant hoppers, with no effect on human metabolism [111, 112]. Lectins act on insects by binding to glycoproteins present in insect gut epithelium, eventually causing death of insect by inhibiting absorption of nutrients. It was believed that N-linked glycans in insects were exclusively of the high mannose type; therefore, there are great interests, especially in mannose-specific plant lectins, as possible insecticidal or insect-deterring molecules for the new pest management strategies [113, 114]. Nevertheless, the lectins possess different sugar specificities and, considering the variety of glycan structures in the bodies of insects, have many different possible targets. Advances have been made in the knowledge related to glycan diversity and function(s) of protein glycosylation in insects, N-glycosylation, and O-glycosylation, and it postulated that the interference in insect glycosylation appears to be a promising strategy for pest insect control [115]. Therefore, it is difficult to predict the exact mode of action of each lectin and even more difficult to understand the variability in insect toxicity upon exposure to different plant lectins. The use of initial bioassays employing artificial diets has led to the most recent advances, such as plant breeding and the construction of fusion proteins, using lectins for targeting the delivery of toxins and to potentiate expected insecticide effects [116–118].

The first lectin known for insecticidal activity was *Galanthus nivalis* agglutinin, which belongs to a superfamily of alpha-D-mannose-specific plant bulb lectins [105, 119]. The mannose-binding lectins have shown strong insecticidal activity against chewing and sap-sucking insects and particularly in controlling aphids [120–124]. Lectin isolated from bulbs of *Phycella australis* presented a strong insecticidal activity against the pea aphid and green peach aphid, affecting the survival, feeding behavior, and fecundity of aphids, where *Acyrthosiphon pisum* proved to be particularly sensitive [125].

No considerable mortality effect of ASA lectins (native or recombinant lectins) was shown on larvae of potato moths (*Tecia solanivora*); however, recombinant ASAII lectin had an effect on the pupa mortality, which was bigger than the native lectin effect. The effect of lectins on the weight and fertility of adults showed that both lectins had a big effect on fertility when the lectin is used in a low concentration (lower than 0.003 mg/mL), and, in some cases, lectins produced malformations in female adults [126]. Fitches et al. found toxic effects on *Acyrthosiphon pisum* using both recombinant lectins; however, ASA II was more toxic than ASA I, at the same dose [127].

Lectins from legume family have shown insectistastic and insecticidal activity [52] (**Table 4**). The lectins from seeds of *Canavalia brasiliensis*, *Dioclea grandiflora*, *Dioclea rostrata*, *Cratylia floribunda*, and *Phaseolus vulgaris* have shown to protect seeds against the beetle *Callosobruchus maculatus*. In general, the plant lectins are the most potent agents against insect pests of a variety of crops including wheat, rice, tobacco, and potatoes [128]. *Canavalia* lectins exhibited a range of different toxicities toward *Artemia nauplii* and bound to a similar area in the digestive tract; differences in spatial arrangement and volume of CRD (carbohydrate recognition domain) may explain the variation of the toxicity showed by each lectin despite the high structural similarity [129]. The sensitivity of different insect species to the insecticidal effects of lectin ingestion is variable, and the binding of a lectin to the gut does not necessarily imply toxicity. Other studies signal that lectins affect various insect hydrolytic enzymes such as glucosidases, phosphatases, and proteases which are involved in digestion, development, growth, and detoxification. To date a great number of studies have shown lectin toxicity in insects belonging to different orders, including Lepidoptera, Coleoptera, and Hemiptera. However, the exact mode of action of lectins in providing resistance against insects remains unclear. The most relevant property of lectin's anti-insect activity can be related to its interactions with different glycoproteins or glycan structures in insects, which may interfere with a number of physiological processes in these organisms. Lectins possess at least one carbohydrate-binding domain and different sugar specificities, possible targets for lectin binding are numerous, and several mechanisms can be associated (**Figure 2**).

Preliminary evidence of Gleheda's insecticidal activity against Colorado potato beetle larvae (*Leptinotarsa decemlineata*) has been obtained using a single dose of lectin [130]; it would have been very interesting to carry out dose-response experiments and to assay several insect pests to elucidate whether the lectin was insect specific. Nevertheless, Gleheda's insecticidal activity stresses the importance of this unusual lectin, begging the question of whether such activity is shared by other Lamiaceae lectins. To date Lamiaceae lectin is unique with known insecticidal activity. The importance of lectins due to their insecticidal properties, isolation of native lectins, and lectin genes could be agronomically important tools for crop plants for developing resistance against insect pests mainly for sap-sucking

Lectin	Insect pests	Activity	References
PSA	*Meligethes aeneus*	Insecticidal, insectistatic	[140]
ConA	*Tarophagous proserpina*	Insectistatic	[141]
Gleheda	*Leptinotarsa decemlineata*	Insectistatic	[130]
ConA	*Callosobruchus maculatus*	Insectistatic	[142]
ConA	*Helicoverpa armigera*	Insectistastic	[143]
GS-II	*Callosobruchus maculatus*	Insectistastic	[144]
PHA	*Callosobruchus maculatus*	Insecticidal	[145]
PHA-E	*Empoasca fabae*	Insecticidal	[146]
Bmoll	*Anagasta kuehniella*	Insecticidal	[147]
	Zabrotes subfasciatus		
	Callosobruchus maculatus		
	Callosobruchus maculatus		
DGL	*C. maculatus*		[108]
DRL			
CFL			

Pisum sativum (PSA), *Canavalia ensiformis* (concanavalin A (ConA)), *Glechoma hederacea* (Gleheda), GS-II: *Griffonia simplicifolia* aglutinina, *Phaseolus vulgaris* (PHA), *Bauhinia monandra* leaf lectin (bmoll), *Dioclea grandiflora* (DGL), *D. rostrata* (DRL), *Cratylia floribunda* (CFL). Taking from Calvacante et al. [60] and modified.

Table 4. Legume lectins domain with insectistatic and insecticidal activity.

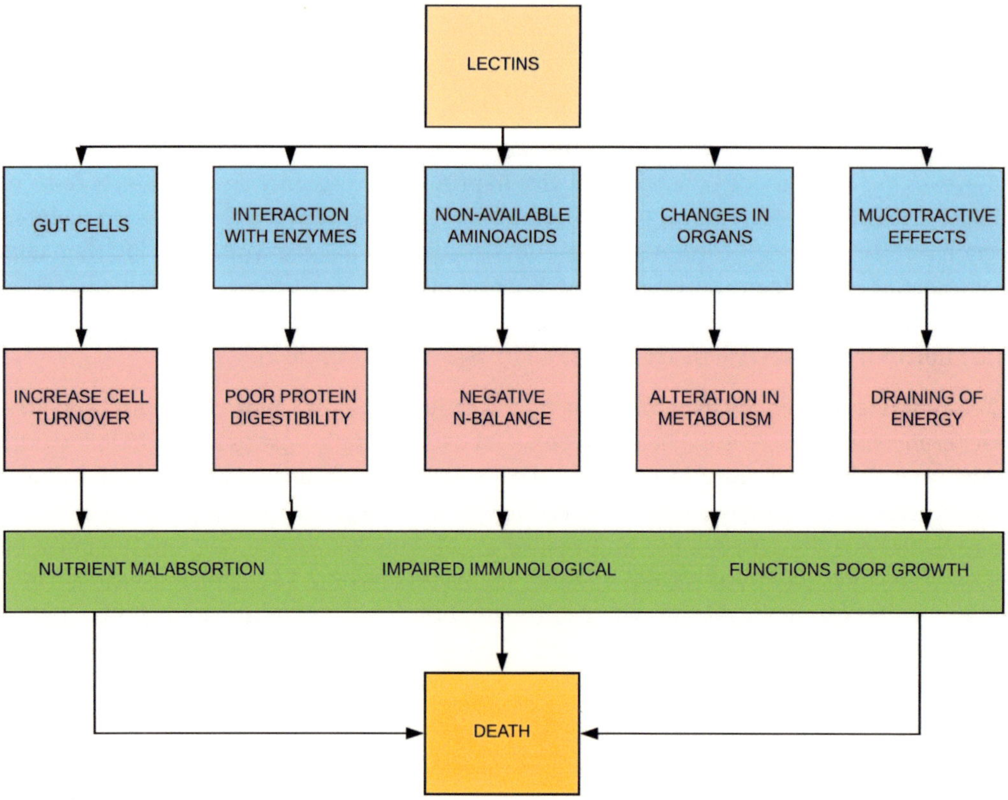

Figure 2. Possible targets and associated mechanisms of lectin anti-insect activity. Lectins have antinutritional properties by which they interact with several targets in digestive tract and other organs.

insect. These proteins are very interesting, and its molecular properties have been well described; however, there is still a long way to study and learn about the mechanisms of these molecules at a physiological and molecular level.

Author details

Edgar Antonio Reyes-Montaño* and Nohora Angélica Vega-Castro

*Address all correspondence to: eareyesm@unal.edu.co

Department of Chemistry, Protein Research Group, National University of Colombia, Bogotá, Colombia

References

[1] Goldstein IJ, Hughes RC, Monsigny M, Osawa T, Sharon N. What should be called a lectin? Nature. 1980;**285**:66

[2] Kumar KK, Lalith Prakash CK, Sumanthi J, Reddy GS, Shekar PC, Reddy B. Biological role of lectins: A review. Journal of Orofacial Sciences. 2012;**4**:20-25

[3] Liu B, Bian HJ, Bao JK. Plant lectins: Potential antineoplastic drugs from bench to clinic. Cancer Letters. 2010;**287**:1-12

[4] Sharon N, Lis H. Lectins as cell recognition molecules. Science. 1989;**246**:227-234

[5] Ghazarian H, Idoni B, Oppenheimer SB. A glycobiology review: carbohydrates, lectins, and implications in cancer therapeutics. Acta Histochemica. 2011;**113**(3):236-247. DOI: 10.1016/j.acthis.2010.02.004

[6] Gorelik E, Galili U, Raz A. On the role of cell surface carbohydrates and their binding proteins (lectins) in tumor metastasis. Cancer Metastasis Reviews. 2001;**20**:245-277

[7] Drickamer K. Two distinct classes of carbohydrate recognition domains in animal lectins. The Journal of Biological Chemistry. 1988;**263**:9557-9560

[8] Singh H, Sarathi SP. Insight of lectins – A review. Journal of Scientific and Engineering Research. 2012;**3**:4

[9] Van Damme EJM, Peumans WJ, Barre A, Rouge P. Plant lectins: A composite of several distinct families of structurally and evolutionary related proteins with diverse biological roles. Critical Reviews in Plant Sciences. 1998;**17**:575-692

[10] Fujimoto Z, Tateno H, Hirabayashi J. Lectin structures: Classification based on the 3-D structures. Lectins: Methods and Protocols. 2014:579-606

[11] Van Damme EJ, Lannoo N, Fouquaert E, Peumans WJ. The identification of inducible cytoplasmic/nuclear carbohydrate-binding proteins urges to develop novel concepts about the role of plant lectins. Glycoconjugate Journal. 2003;**20**:449-460

[12] Macedo MLR, Oliveira CF, Oliveira CT. Insecticidal activity of plant lectins and potential application in crop protection. Molecules. 2015;**20**(2):2014-2033

[13] Sharon N, Lis H. Legume lectins—A large family of homologous proteins. The FASEB Journal. 1990;**4**:3198-3208

[14] Young NM, Oomen RP. Analysis of sequence variation among legume lectins: A ring of hypervariable residues forms the perimeter of the carbohydrate-binding site. Journal of Molecular Biology. 1992;**228**:924-934

[15] Peumans WJ, Van Damme EJM. Lectins as plant defense proteins. Plant Physiology. 1995;**109**:347-352. DOI: 10.1104/pp.109.2.347

[16] Lannoo N, Van Damme EJ. Lectin domains at the frontiers of plant defense. Frontiers in Plant Science. 2014;**5**:1-16

[17] Wang H, Liu W, Ng T, Ooi V, Chang S. The immunomodulatory and antitumor activities of lectins from the mushroom Tricholoma mongolicum. Immunopharmacology. 1996;**31**(3):205-211

[18] Majee S, Biswas G. Exploring plant lectins in diagnosis, prophylaxis and therapy. Journal of Medicinal Plants Research. 2013;**7**(47):3444-3451

[19] Ashraf M, Khan R. Mitogenic lectins. Medical Science Monitor. 2003;**9**(11):RA265-RA269

[20] He X, Ji N, Xiang X, Luo P, Bao J. Purification, characterization, and molecular cloning of a novel antifungal lectin from the roots of *Ophioglossum pedunculosum*. Applied Biochemistry and Biotechnology. 2011;**165**(7):1458-1472

[21] De Albuquerque L, De S'a Santana G, Napoleao T, Coelho L, Da Silva M, Paiva P. Antifungal activity of Microgramma vacciniifolia rhizome lectin on genetically distinct *Fusarium oxysporum* f. sp. Lycopersiciraces. Applied Biochemistry and Biotechnology. 2014;**172**(2):1098-1105

[22] Yao Q, Wu C, Luo P. A new chitin-binding lectin from rhizome of *Setcreasea purpurea* with antifungal, antiviral and apoptosis-inducing activities. Process Biochemistry. 2010;**45**(9): 1477-1485

[23] Trindade M, Lopes J, Soares-Costa A. Structural characterization of novel chitin-binding lectins from the genus Artocarpus and their antifungal activity. Biochimica et Biophysica Acta—Proteins and Proteomics. 2006;**1764**(1):146-152

[24] Afonso-Cardoso S, Silva C, Ferreira M, Souza M. Effect of the *Synadenium carinatum* latex lectin (ScLL) on Leishmania (Leishmania) amazonensis infection in murine macrophages. Experimental Parasitology. 2011;**128**(1):61-67

[25] Albuquerque L, Pontual E, Santana G. Toxic effects of *Microgramma vacciniifolia* rhizome lectin on *Artemia salina*, human cells, and the schistosomiasis vector *Biomphalaria glabrata*. Acta Tropica. 2014;**138**:23-27

[26] Santos A, Cavada B, Rocha B, Nascimento K, SantAna A. Toxicity of some glucose/mannose-binding lectins to Biomphalaria glabrata and Artemia salina. Bioresource Technology. 2010;**101**(2):794-798

[27] Batterbury M, Tebbs C, Rhodes J, Grierson I. Agaricus bisporus (edible mushroom lectin) inhibits ocular fibroblast proliferation and collagen lattice contraction. Experimental Eye Research. 2002;**74**(3):361-370

[28] Yu L, Fernig D, Smith J, Milton J, Rhodes J. Reversible inhibition of proliferation of epithelial cell lines by agaricus bisporus (edible mushroom) lectin. Cancer Research. 1993;**53**(19):4627-4632

[29] Pereira P, Del Aguila E, Verlcimo M, Zingali R, Paschoalin M, Silva J. Purification and characterization of the lectin from taro (Colocasia esculenta) and its effect on mouse splenocyte proliferation in vitro and in vivo. Protein Journal. 2014;**33**(1):92-99

[30] da Silva L, Filho C, De Paula R, Coelho L, Da Silva M, Correia M. Cratylia mollis lectin: A versatile tool for biomedical studies. Current Bioactive Compounds. 2014;**10**(1):44-54

[31] Jacinto A, Cordeiro J. The role of transcription independent damage signals in the initiation of epithelial wound healing. Nature Reviews Molecular Cell Biology. 2013;**14**(4):249-262

[32] Velnar T, Bailey T, Smrkolj V. The wound healing process: An overview of the cellular and molecular mechanisms. Journal of International Medical Research. 2009;**37**(5): 1528-1542

[33] Coriolano M, De Melo C, Silva F. Parkia pendula seed lectin: Potential use to treat cutaneous wounds in healthy and immunocompromised mice. Applied Biochemistry and Biotechnology. 2014;**172**(5):2682-2693

[34] Sheng Y, He H, Zou H. Poly(lactic acid) nanoparticles coated with combined WGA and water-soluble chitosan for mucosal delivery of *β*-galactosidase. Drug Delivery. 2014;**21**(5):370-378

[35] Jain S, Gupta M, Sahoo A, Pandey A, Jain A. Lectin conjugated gastro-retentive microspheres of amoxicillin for effective treatment of helicobacter pylori. Current Science. 2014;**106**(2):267-276

[36] Adebisi A, Conway B. Lectin-conjugated microspheres for eradication of *Helicobacter pylori* infection and interaction with mucus. International Journal of Pharmaceutics. 2014;**470**(1):28-40

[37] Schnegelsberg B, Schumacher U, Valentiner U. Lectin histochemistry of metastasizing and non-metastasizing breast and colon cancer cells. Anticancer Research. 2011;**31**(5):1589-1597

[38] Beltrao E, Medeiros P, Rodrigues O. Parkia pendula lectin as histochemistry marker for meningothelial tumour. European Journal of Histochemistry. 2003;**47**(2):139-142

[39] Chandler K, Goldman R. Glycoprotein disease markers and single protein-omics. Molecular and Cellular Proteomics. 2013;**12**(4):836-845

[40] Li F, Feng Y, Yang L, Li L, Tang C, Tang B. A selective novel non-enzyme glucose amperometric biosensor based on lectin-sugar binding on thionine modified electrode. Biosensors and Bioelectronics. 2011;**26**(5):2489-2494

[41] Luna D, Oliveira M, Nogueira M, Andrade C. Biosensor based on lectin and lipid membranes for detection of serum glycoproteins in infected patients with dengue. Chemistry and Physics of Lipids. 2014;**180**:7-14

[42] De Oliveira A, Silva L, Lima T. Biotechnological value of Moringa oleifera seed cake as source of insecticidal lectin against Aedes aegypti. Process Biochemistry. 2016;**51**(10):1683-1690

[43] Williams C, Collier C, Nemacheck J, Liang C, Cambron S. A lectin-like wheat gene responds systemically to attempted feeding by avirulent first-instar hessian fly larvae. Journal of Chemical Ecology. 2002;**28**:1411-1428

[44] Puthoff DP, Sardesai N, Subramanyam S, Nemacheck JA, Williams CE. Hfr-2, a wheat cytolytic toxin-like gene, is up-regulated by virulent hessian fly larval feeding. Molecular Plant Pathology. 2005;**6**:411-423

[45] Pyati P, Chellamuthu A, Gatehouse AMR, Fitches E, Gatehouse JA. Insecticidal activity of wheat hessian fly responsive proteins HFR-1 and HFR-3 towards a non-target wheat pest, cereal aphid (*Sitobion avenae* F.). Journal of Insect Physiology. 2012;**58**:991-999

[46] Vandenborre G, Groten K, Smagghe G, Lannoo N, Baldwin IT, van Damme EJM. Nicotiana tabacum agglutinin is active against Lepidopteran pest insects. Journal of Experimental Botany. 2010;**61**:1003-1014

[47] Sharon N, Lis H. Lectins. Netherlans: Springer Science; 2007

[48] Rüdiger H, Gabius H. The history of lectinology. In: Gabius HJ, editor. The Sugar Code. Fundamentals of Glycosciences. Weinheim, Germany: Wiley; 2009. pp. 261-268

[49] De Schutter K, Van Damme EJM. Protein-carbohydrate interactions as part of plant Defense and animal immunity. Molecules. 2015;**20**:9029-9053

[50] Dang L, Van Damme EJM. Review toxic proteins in plants. Phytochemistry. 2015;**117**: 51-64

[51] Lima TE, Sartori ALB, Rodrigues MLM. Plant antiherbivore defenses in Fabaceae species of the Chaco. Brazilian Journal of Biology. 2017;**77**(2):299-303. DOI: 10.1590/1519-6984.12815

[52] Casas ZY, Reyes-Montaño EA, Vega NA. Lectinas con dominio de Leguminosa: Características estructurales y utilidad como agentes insectistáticos e insecticidas. Chilean. Journal of Agricultural and Animal Sciences (ex Agro-Ciencia). 2016;**32**(2):157-169

[53] Gabius HJ, André S, Kaltner H, Siebert HC. The sugar code: Functional lectinomics. Biochimica et Biophysica Acta (BBA) – General Subjects. 2002;**1572**(2-3):165-177

[54] Maciel EVM, Araújo-Filho VS, Nakazawa M, Gomes YM, Coelho LCBB, Correia MTS. Mitogenic activity of Cratylia mollis lectin on human lymphocytes. Biologicals. 2004;**32**:57-60

[55] Dan X, Ng TB. Two legume defense proteins suppress the mobility of nasopharyngeal carcinoma cells. Journal of Enzyme Inhibition and Medicinal Chemistry. 2016;**31**(6):1328-1334

[56] Rodríguez-Navarro DN, Dardanelli MS, Ruiz-Sainz JE. Attachment of bacteria to the roots of higher plants. FEMS Microbiology Letters. 2007;**272**:127-136. DOI: 10.1111/j.1574-6968.2007.00761.x

[57] Albareda M, Dardanelli M, Sousa C, megias M, Temprano F, Rodriguwz-Navarro D. Factors affecting the attachment of rhizospheric bacteria to bean and soybean roots. FEMS Microbiology Letters. 2006;**259**(1):67-73 https://doi.org/10.1111/j.1574-6968.2006.00244.x

[58] Brewina N, more KIS. Legume lectins and nodulation by Rhizobium. Trends in Plant Science. 1997;**2**(3):92-98. DOI: 10.1016/S1360-1385(96)10058-3

[59] De HoV P, Brill L, Hirsch A. Plant lectins: The ties that bind in root symbiosis and plant defense. Molecular Genetics and Genomics. 2009;**282**:1-15. DOI: 10.1007/s00438-009-0460-8

[60] Calvacante T, Firmino N, Solon F, de Andrade C, Albuquerque R. Plant lectins as alternative tools against bacterial biofilms. American Journal of Microbiological Research. 2014;**8**(27):2555-2564. DOI:105897/AJMR2014.6710

[61] Vasconcelos M, Vassiliepe F, Alves V, Colares H,1 Santiago Nascimento K, Holanda A, Cavada B, Holanda E, Henriques M and Pereira M. Effect of algae and plant lectins on planktonic growth and biofilm formation in clinically relevant bacteria and yeasts. BioMed Research International. 2014:1-9. Article ID: 365272. DOI: 10.1155/2014/365272

[62] Syed P, Gidwani K, Kekki H, Leivo J, Pettersson K, Lamminmaki U. Role of lectin microarrays in cancer diagnosis. Proteomics. 2016;**16**(8):1257-1265

[63] Gorakshakar AC, Ghosh K. Use of lectins in immunohematology. Asian Journal of Transfusion Science. 2016;**10**(1):12-21

[64] Coulibaly F, Youan B. Current status of lectin-based cancer diagnosis and therapy. AIMS Molecular Science. 2017;**4**(1):1-27. DOI: 10.3934/molsci.2017.1.1

[65] Lagarda-Diaz I, Guzman-Partida A, Vazquez-Moreno L. Review legume lectins: Proteins with diverse applications. International Journal of Molecular Sciences. 2017;**18**:1242. DOI: 10.3390/ijms18061242

[66] Poiroux G, Barre A, Van Damme E, Benoist H, Rougé P. Review plant lectins targeting O-glycans at the cell surface as tools for cancer diagnosis, prognosis and therapy. International Journal of Molecular Sciences. 2017;**18**(6):1232. DOI: 10.3390/ijms18061232

[67] Goldstein IJ, Poretz RD., Isolation physicochemical characterization and carbohydrate-binding specificity of lectins. In: Liener IE, Sharon N, Goldstein IJ, editors. The Lectins Properties, Functions and Applications in Biology and Medicine. Elsevier; 1986. pp. 233-247

[68] Loris R. Principles of structures of animal and plant lectins. Biochimica et Biophysica Acta. 2002;**1572**(2-3):198-208

[69] De Queiroz LP, Pastore JFB, Cardoso D, Snak C de C, Lima AL de C, Gagnon E, Vatanparast M, Holland AE, Egan AN. A multilocus phylogenetic analysis reveals the monophyly of a recircumscribed papilionoid legume tribe Diocleae with well-supported generic relationships. Molecular Phylogenetics and Evolution. 2015;**90**:1-19

[70] Moreira RA, Barros ACH, Stewart JC, Pusztai A. Isolation and characterization of a lectin from the seeds of *Dioclea grandiflora* (Mart.). Planta (Heidelh.). 1983;**158**:63-69

[71] Richardson M, Campos FDAP, Moreira RA, Ainouz IL, Begbie R, Watt WB, Pusztai A. The complete amino acid sequence of the major alpha subunit of the lectin from the seeds of Dioclea grandiflora (Mart). European Journal of Biochemistry. 1984;**144**:101-111

[72] Pérez G, Hernández M, Mora E. A lectin from the seeds of Dioclea lehmanni. Phytochemistry. 1990;**29**:1745-1749

[73] Sierra AY, Pérez GE. purificación y caracterización de dos lectinas en semillas de Dioclea sericea. Revista de la Academia Colombiana de Ciencias. 1999;**23**(88):445-454

[74] Van Damme EJM, Peumans WJ, Pusztai A, Bardocz S. Handbook of Plant Lectins: Properties and Biomedical Applications. Chichester, UK: John Wiley; 1998

[75] Pérez G. Isolation and characterization of a novel lectin from Dioclea lehmanni (Fabaceae) seeds. The International Journal of Biochemistry & Cell Biology. 1998;**30**:843-853

[76] Melgarejo LM, Vega N, Pérez G. Isolation and characterization of novel lectins from Canavalia ensiformis DC and Dioclea grandiflora Mart. Ex Benth. seeds. Brazilian Journal of Plant Physiology. 2005;**17**(3):315-324

[77] Quintero M. Elucidación parcial de la estructura primaria de la lectina LGL-P2 y purificación y caracterización parcial de la lectina LGL-P4 presentes en semillas de *Galactia lindenii.* Trabajo de grado de Maestría, Departamento de Química: Universidad Nacional de Colombia; 2014

[78] Melgarejo LM, Pérez G. Immunolocalization of the lectins p2 Y p4 from Dioclea lehmanni seeds. Plant & Cell Physiology. 1997;**38**(4):480-483

[79] Nagano C, Calvete J, Barettino D, Perez A, Cavada B, Sanz L. Insights into the structural basis of the pH-dependent dimer–tetramer equilibrium through crystallographic analysis of recombinant Diocleinae lectins. The Biochemical Journal. 2008;**409**:417-428. DOI: 10.1042/BJ20070942

[80] Cavada B, Barbosa T, Arruda S, Grangeiro T, Barral-Netto M. Revisiting proteus: Do minor changes in lectin structure matter in biological activity? Lessons from and potential biotechnological uses of the Diocleinae Subtribu lectins. Current Protein & Peptide Science. 2001;**2**:123-135

[81] Pérez G, Vega N. Lamiaceae lectins. Functional Plant Science and Biotechnology 2007;**1**(2):288-299

[82] Springer GF. T and Tn, general carcinoma autoantigens. Science. 1984;**224**(4654):1198-1206

[83] Lisowska E. Tn Antigens and their significance in oncology. Acta Biochimica Polonica. 1995;**42**(1):11-17

[84] Berger EG. Tn-syndrome. Biochimica et Biophysica Acta (BBA) – Molecular Basis of Disease. 1999;**1455**(2-3):255-268

[85] Ju T, Wang Y, Aryal RP, Lehoux SD, Ding X, Kudelka MR, Cutler C, Zeng J, Wang J, Sun X, Heimburg-Molinaro J, Smith DF, Cummings RD. Tn and sialyl-Tn antigens, aberrant O-glycomics as human disease markers. Proteomics. Clinical Applications. 2013;**7**(9-10):618-631. DOI: 10.1002/prca.201300024

[86] Ju T, Otto VI, Cummings RD. The Tn antigen-structural simplicity and biological complexity. Angewandte Chemie International Edition in English. 2011;**50**:1770-1791

[87] Li Q, Anver M, Butcher D, Gildersleeve J. Resolving conflicting data on expression of the Tn antigen and implications for clinical trials with cancer vaccines. Molecular Cancer Therapeutics. 2009;**8**(4):971-979. DOI: 10.1158/1535-7163.MCT-08-0934

[88] Cazet A, Julien S, Bobowski M, Burchell J, Delannoy P. Tumour-associated carbohydrate antigens in breast cancer. Breast Cancer Research. 2010;**12**:204

[89] Ju T, Aryal RP, Kudelka MR, Wang Y, Cummings RD. The Cosmc connection to the Tn antigen in cancer. Cancer Biomarkers. 2014;**14**(1):63-81. DOI: 10.3233/CBM-130375

[90] Fu C, Zhao H, Wang Y, Cai H, Xiao Y, Zeng Y, Chen H. Tumor-associated antigens: Tn antigen, sTn antigen, and T antigen. HLA Journal. 2016;**88**(6):275-286. DOI: 10.1111/tan.12900

[91] Review MJ. The role of Sialyl-Tn en cancer. International Journal of Molecular Sciences. 2016;**17**(3):275

[92] Osinaga E, Bay S, Tello D, Babino A, Pritsch O, Assemat CD, Nakada H, Alazari P. Analysis of the fine specificity of Tn-binding proteins using synthetic glycopeptide epitopes and a biosensor based on surface plasmon resonance spectroscopy. FEBS Letters. 2000;**469**:24-28

[93] Kitagaki-Ogawa H, Matsumoto I, Seno N, Takahashi N, Endo S, Arata Y. Characterization of the carbohydrate moiety of Clerodendron trichotomum lectins: Its structure and reactivity toward plant lectins. European Journal of Biochemistry. 1986;**161**:779-785

[94] Medeiros A, Bianchi S, Calvete JJ, Balter H, Bay S, Robles A, et al. Biochemical and functional characterization of the Tn-specific lectin from Salvia sclarea seeds. European Journal of Biochemistry. 2000;**267**:1434-1434

[95] Lis H, Latter H, Adar R, Sharon N. Isolation of two blood type A and N specific isolectins from Moluccella laevis seeds. FEBS Letters. 1988;**233**:191-195

[96] Wang W, Peumans WJ, Rougé P, Rossi C, Proost P, Chen J, et al. Leaves of the Lamiaceae species Glechoma hederacea (ground ivy) contain a lectin that is structurally and evolutionary related to the legume lectins. The Plant Journal. 2003;**33**:239-304

[97] Wenping H, Limin H, Zhezhi W. Molecular cloning and expression of a novel gene related to legume lectin from Salvia miltiorrhiza Bunge. African Journal of Biotechnology. 2015;**14**(28):2234-2243

[98] Pérez G, Vega N, Fernández-Alonso JL. Prospección de lectinas en especies de labiadas colombianas. Un enfoque sistemático-ecológico-II. Caldasia. 2006;**28**:179-195

[99] Fernández-Alonso JL, Vega N, Pérez G. Lectin prospecting in Colombian Labiatae. A systematic ecological approach – III. Mainily exotic species (cultivated or naturalised). Caldasia. 2009;**31**:227-245

[100] Filgueira-Duarte JJ, Pérez G. Producción de lectinas Tn-específicas obtenidas de Salvia palifolia y Hyptis mutabilis por variación somaclonal celular. Revista Facultad De Ciencias Básicas, Universidad Militar Nueva Granada. 2013;**9**:134-141

[101] Vega N, Pérez G. Isolation and characterisation of a Salvia bogotensis seed lectin specific for the Tn antigen. Phytochemistry. 2006;**67**(4):347-355

[102] Wilches-Torres A, Rojas-Caraballo J, Sanabria E, Reyes-Montaño E, Fernández-Alonso JL, Varrot A, Imberty A, Vega N. Purification and biochemical characterization of a T/Tn specific lectin from *Lepechinia bullata* Seeds (*Lamiaceae*). International Journal of Pharmacy and Pharmaceutical Sciences. 2017;**9**(11):165-174

[103] Rougé P, Peumans WJ, Van Damme EJM, Barre A, Singh T, Wu JH, Wu AM. Structure-function relationships of plant lectins that specifically recognize T and Tn antigens. In: Wu AM, editor. The Molecular Immunology of Complex Carbohydrates. 3rd ed. Springer; 2011. pp. 157-170

[104] Wilches A. Aproximación a la estructura primaria de lectinas específicas Para el antígeno Tn e identificación de nuevas lectinas específicas Para glucosa/manosa en

Semillas de *Salvia bogotensis* y *Lepechinia bullata*. Approach primary structure of Tn antigen specific lectins and identification of new glucose/mannose specific lectins in *Salvia bogotensis* and *Lepechinia bullata* seeds [doctoral dissertation]. Sciences Faculty National University from Colombia. 2017

[105] Van Damme EJM, Allen AK, Peumans WJ. Isolation and characterization of a lectin with exclusive specificity towards mannose from snowdrop (*Galanthus nivalis*) bulbs. FEBS Letters. 1987;**215**:140-114

[106] De Vasconcelos MA, Alves AC, Carneiro RF, Sampaio Dias AH, Viana Martins FW, Batista Cajazeiras J, Nagano CS, Holanda Teixeira E, do Nascimento KS, Cavada BS. Purification and primary structure of a novel mannose-specific lectin from *Centrolobium microchaete* Mart seeds. International Journal of Biological Macromolecules. 2015;**81**:600-607

[107] Barre A, Bourne Y, Van Damme EJ. Mannose-binding plant lectins: Different structural scaffolds for a common sugar-recognition process. Biochimie. 2001;**83**(7):645-651

[108] Gatehouse AMR, Powell KS, Peumans WJ, Van Damme EJM, Gatehouse JA. Insecticidal properties of plant lectins: Their potential in plant protection. In: Pusztai A, Bardocz S, editors. Lectins: Biomedical Perspectives. London: Taylor & Francis; 1995. pp. 35-58

[109] Vandenborre G, Smagghe G, Van Damme EJ. Plant lectins as defense proteins against phytophagous insects. Phytochemistry. 2011;**72**(13):1538-1550

[110] Jaber K, Haubruge E, Francis F. Development of entomotoxic molecules as control agents: Illustration of some protein potential uses and limits of lectins. Biotechnologie, Agronomie, Société et Environnement. 2010;**14**(1):225-241

[111] Karimi J, Allahyari M, Bandani AR. Lectins and their roles in pests control. In: Bandani AR, editor. New Perspectives in Plant Protection. InTech; 2012. DOI: 10.5772/39377

[112] Michiels K, Van Damme EJM, Smagghe G. Plant-insect interactions: What can we learn from plant lectins? Archives of Insect Biochemistry and Physiology. 2010;**73**:193-221. DOI: 10.1002/arch.20351

[113] Vandenborre G, Smagghe G, Ghesquie're B, Menschaert G, Nagender Rao R, Gevaert K, Van Damme EJM. Diversity in protein glycosylation among insect species. PLoS One. 2011;**6**(2):e16682. DOI: 10.1371/journal.pone.0016682

[114] Hamshou M, Van Damme EJM, Vandenborre G, Ghesquiere B, Trooskens G, Gevaert K, Smagghe G. GalNAc/Gal-binding Rhizoctonia solani Agglutinin has antiproliferative activity in *Drosophila melanogaster* S2 cells via MAPK and JAK/STAT signaling. PLoS One. 2012;**7**(4):e33680. DOI: 10.1371/journal.pone.0033680

[115] Walski T, De Schutter K, Van Damme E, Smagghe G. Diversity and functions of protein glycosylation in insects. Insect Biochemistry and Molecular Biology. 2017;**83**:21-34

[116] Fitches EC, Pyati P, King GF, Gatehouse JA. Fusion to snowdrop Lectin magnifies the oral activity of insecticidal v-Hexatoxin-Hv1a peptide by enabling its delivery to the central nervous system. PLoS One. 2012;**7**(6):e39389. DOI: 10.1371/journal.pone.0039389

[117] Yang S, Pyati P, Fitches E, Gatehouse J. A recombinant fusion protein containing a spider toxin specific for the insect voltage-gated sodium ion channel shows oral toxicity towards insects of different orders. Insect Biochemistry and Molecular Biology. 2014;**47**(100):1-11. DOI: 10.1016/j.ibmb.2014.01.007

[118] Rani S, Sharma V, Hada A, Bhattacharya RC, Koundal KR. Fusion gene construct preparation with lectin and protease inhibitor genes against aphids and efficient genetic transformation of Brassica juncea using cotyledons explants. Acta Physiologiae Plantarum. 2017;**39**(5):1-13

[119] Stoger E, Williams S, Christou P, Down R, Gatehouse J. Expression of the insecticidal lectin from snowdrop (*Galanthus nivalis* agglutinin; GNA) in transgenic wheat plants: Effects on predation by the grain aphid Sitobion avenae. Molecular Breeding. 1999;**5**(1):65-73

[120] Sauvion N, Nardon C, Febvay G, Gatehouse AM, Rahbe Y. Binding of the insecticidal lectin Concanavalin A in pea aphid, *Acyrthosiphon pisum* (Harris) and induced effects on the structure of midgut epithelial cells. Journal of Insect Physiology. 2004;**50**(12):1137-1150

[121] Sadeghi A, Van Damme EJM, Michiels K, Kabera A, Smagghe G. Acute and chronic insecticidal activity of a new mannose-binding lectin from Allium porrum against Acyrthosiphon pisum via an artificial diet. The Canadian Entomologist. 2009;**141**:95-101

[122] Das A, Roy A, Hess D, Das S. Characterization of a highly potent insecticidal lectin from Colocasia esculenta tuber and cloning of its coding sequence. American Journal of Plant Sciences. 2013;**4**:408-416 http://dx.doi.org/10.4236/ajps.2013.42A053

[123] Roy A, Das S. Molecular mechanism underlying the entomotoxic effect of Colocasia esculenta tuber agglutinin against *Dysdercus cingulatus*. Insects. 2015;**6**:827-846. DOI: 10.3390/insects6040827

[124] Sprawka I, Goławska S, Parzych T, Sytykiewicz H, Czerniewicz P. Apoptosis induction by Concanavalin A in gut cells of grain aphid. Arthropod-Plant Interactions. 2015;**9**:133-140. DOI: 10.1007/s11829-015-9356-1

[125] Zapata N, Van Damme EJM, Vargas M, Devotto L, Smagghe G. Insecticidal activity of a protein extracted from bulbs of Phycella australis Ravenna against the aphids *Acyrthosiphon pisum* Harris and Myzus persicae Sulzer. Chilean Journal of Agricultural Research. 2016;**76**(2):188-194

[126] Carreño N. Evaluación preliminar de actividad insecticida de lectina ASA II a partir de ajo (*Allium sativum*), sobre Tecia solanivora. Preliminary evaluation of the insecticidal activity of lectin ASA II on tecia solanivora [Master dissertation]. Sciences Faculty. National University from Colombia. 2013

[127] Fitches E, Wiles D, Douglas A, Hinchliffe G, Audsley N, Gatehouse J. The insecticidal activity of recombinant garlic lectins toward aphids. Insect Biochemistry and Molecular Biology. 2008;**38**:905-915

[128] Sousa Arruda F, Alves Melo A, Alves Vasconcelos M, Farias Carneiro R, Barroso-Neto I, Silva S, Nascimento Pereira Jr F, Shiniti Nagano C, Santiago Nascimento K,

Holanda Teixeira E, Saker-Sampaio S, Sousa Cavada B, Holanda Sampaio A. Toxicity and binding profile of lectins from the genus Canavalia on brine shrimp. BioMed Research International. 2013:1-7. Article ID: 154542. DOI: 10.1155/2013/154542

[129] Sprawka I, Goławska S, Goławski A, Chrzanowski G, Czerniewicz P, Sytykiewicz H. Entomotoxic action of jackbean lectin (Con A) in bird cherry-oat aphid through the effect on insect enzymes, Journal of Plant Interactions. 2014;**9**:(1):425-433. DOI: 10.1080/17429145.2013.848947

[130] Wang W, Hause B, Peumans WJ, Smagghe G, Mackie A, Fraser R, Van Damme EJM. The Tn antigen-specific lectin from ground ivy is an insecticidal protein with an unusual physiology. Plant Physiology. 2003;**132**(3):1322-1334

[131] Moreira R, Monteiro A, Oliveira J, Sousa-Cavada B. Isolation and characterization of *Dioclea altíssima* var. megacarpa seed lectin. Phytochemistry. 1997;**46**:139-144

[132] Moreira RA, Cordeiro EF, Ramos MV, Grangeiro TB, Martins JL, Oliveira JTA, et al. Isolation and partial characterization of a lectin from seeds of Dioclea violacea. Revista Brasileira de Fisiologia Vegetal. 1996;**8**:23-29

[133] Pinto Jr VR, Queiroz de Santiago M, da Silva Osterne VJ, Almeida Correia JL, Nascimento Pereira JF, Batista Cajazeiras J, de Vasconcelos MA, Teixeira EH, do Nascimento AS, Miguel TB, Miguel Ede C, Sampaio AH, do Nascimento KS, Nagano CS, Cavada BS. Purification, partial characterization and immobilization of a mannose-specific lectin from seeds of *Dioclea lasiophylla* Mart. Molecules. 2013;**18**(9): 10857-10869. DOI: 10.3390/molecules180910857

[134] Correia JLA, Nascimento ASF, Cajazeiras JB, Gondim ACS, Pereira RI, Sousa BL, Silva ALC, Garcia W, Teixeira EH, Nascimento KS, Rocha BAM, Nagano CS, Sampaio AH, Cavada BS. Molecular characterization and tandem mass spectrometry of the lectin extracted from the seeds of *Dioclea sclerocarpa* Ducke. Molecules. 2011;**16**:9077-9089

[135] De Souza GA, Oliveira PSL, Trapani S, Santos ACO, Rosa JC, Laure HJ, Faça VM, Correia MT, Tavares GA, Oliva G, Coelho LC, Greene LJ. Amino acid sequence and tertiary structure of *Cratylia mollis* seed lectin. Glycobiol. 2003;**13**:961-972

[136] Cavada BS, Marinho E, Souza E, Benevides R, Delatorre P, Souza L, Nascimento KS, Sampaio A, Moreno F, Rustigel J, Canduri F, Azevedo W, Debray H. Purification, partial characterization and preliminary X-ray diffraction analysis of a mannose-specific lectin from *Cymbosema roseum* seeds. Acta Cryst. 2006;**F62**:235-237

[137] Almanza AM, Vega NA, Pérez G. Isolation and characterization of a lectin from Galactia lindenii seeds that recognizes blood group H determinants. Archives of Biochemistry and Biophysics. 2004;**492**:180-190

[138] Rocha B, Moreno F, Delatorre P, Souza E, Marinho E, Benevides R, Rodrigues J, Souza L, Nagano C, Debray H, Sampaio A, De Azevedo W, Cavada B. Purification, characterization, and preliminary X-ray diffraction analysis of a lactose-specific lectin from *Cymbosema roseum* seeds. Applied Biochemistry and Biotechnology. 2009;**152**:383-393

[139] Batista FAH, Goto LS, Garcia W, de Moraes DI, de Oliveira Neto M, Polikarpov I, Cominetti MR, Selistre-de-Araújo HS, Beltramini LM, Ulian Araújo AP.. Camptosemin, a tetrameric lectin of Camptosema ellipticum: structural and functional analysis. European Biophysics Journal. 2010;**39**(8):1193 (205)

[140] Melander M, Ahman I, Kamnert I, Stromdahl AC. Pea lectin expressed transgenically in oilseed rape reduces growth rate of pollen beetle larvae. Transgenic Research. 2003;**12**(5):555-567

[141] Powell KS. Antimetabolic effects of plant lectins towards nymphal stages of the planthoppers Tarophagous proserpina and Nilaparvata lugens. Entomologia Experimentalis et Applicata. 2001;**99**(1):71-78

[142] Sadeghi A, Van Damme EJ, Peumans WJ, Smagghe G. Deterrent activity of plant lectins on cowpea weevil *Callosobruchus maculatus* (F.) oviposition. Phytochemistry. 2006;**67**(18):2078-2084

[143] Shukla SA, Arora R, Sharma HC. Biological activity of soybean trypsin inhibitor and plant lectins against cotton bollworm/legume pod borer, *Helicoverpa armigera* Plant Biotechnology. 2005;**22**(1):1-6

[144] Zhu K, Huesing JE, Shade RE, Bressan RA, Hasegawa PM, Murdock LL. An insecticidal N-acetylglucosamine-specific lectin gene from *Griffonia simplicifolia* (Leguminosae). Plant Physiology. 1996;**110**(1):195-202

[145] Gatehouse AMR, Dewey FM, Dove J, Fenton KA, Pusztai A. Effect of seed lectins from *Phaseolus vulgaris* on the development of larvae of *Callosobruchus maculatus*; mechanism of toxicity. Journal of the Science of Food and Agriculture. 1984;**35**(4):373-380

[146] Habibi J, Backus EA, Huesing JE. Effects of phytohemagglutinin (PHA) on the structure of midgut epithelial cells and localization of its binding sites in western tarnished plant bug, *Lygus hesperus* Knight. Journal of Insect Physiology. 2000;**46**(5):611-619

[147] Macedo MLR, das Graças Machado Freire, M, da Silva MBR, Coelho LCBB. Insecticidal action of Bauhinia monandra leaf lectin (bmoll) against Anagasta kuehniella (lepidoptera: Pyralidae), Zabrotes subfasciatus and *Callosobruchus maculatus* (coleoptera: Bruchidae). Comparative Biochemistry and Physiology. Part A, Molecular & Integrative Physiology. 2007;**146**:486-498

Chapter 3

Role of the Formulation in the Efficacy and Dissipation of Agricultural Insecticides

Karina Buzzetti

Additional information is available at the end of the chapter

http://dx.doi.org/10.5772/intechopen.72340

Abstract

Considering the implications the formulation may have on the effectiveness and residuality of an active ingredient, four trials were conducted comparing two commercial formulations of the diazinon insecticide, two of acetamiprid, two of lambda-cyhalothrin, and, finally, three formulations of imidacloprid. For diazinon and acetamiprid, the comparison parameters used correspond to efficacy against three key pests in apple trees: *Cydia pomonella*, *Diaspidiotus perniciosus*, and *Pseudococcus viburni*; for l-cyhalothrin, efficacy against *C. pomonella* was compared; and for imidacloprid, differences in control *P. viburni* were established. In all cases, their persistence was established in terms of initial and final residue levels in samples of fruits, at 1 and 25 days after application (DAA). Different formulations of the same insecticide correspond to a relevant factor in the general behavior that each product presents in field conditions, being able to affect parameters such as its persistence in the fruit and/or initial deposit of the active ingredient. This variation was demonstrated in the comparison performed on acetamiprid, imidacloprid, and diazinon, but it was not so in l-cyhalothrin. Efficacy was affected in all parameters evaluated for each group of insecticides, demonstrating that different formulations can deliver different biological activity in the control of various pests.

Keywords: pesticide formulations, efficacy, dissipation, residues, insecticide

1. Introduction

The use of multiple crop protection chemicals is a common practice in fruit production, given the requirements of different markets such as the search for plant health, organoleptic quality, and higher yields. In this context, pesticides are applied to agricultural systems for the purpose of protecting plants from damage due to weeds, insects, or diseases [1]. Then, the term pesticide or agrochemical is used to define a wide range of compounds including insecticides, fungicides, herbicides, rodenticides, molluscicides, nematicides, plant growth regulators, defoliant, and others [2].

The role of pesticides in global agriculture has been questioned by United States Environmental Protection Agency (EPA), the European Community, and institutions focused on the consequences of pesticides in human health and environmental subjects [3, 4]. The continuous revaluation of registered pesticides combined with major restrictions like lesser tolerance to residues of pesticides on food has led to an overall trend of reduced risk from pesticides using, for example, innovations in the development of new formulations [5]. It is understood by new formulation a new way of presenting the pesticide for sale, which generally includes, in addition to the active ingredien(s), different adjuvant(s), and/or other formulants combined to render the product useful and effective for the purpose claimed [6].

The supply of plant protection products is wide, although it varies from country to country according to its internal regulations and requirements. However, global development makes it possible to commercially find the same active ingredient formulated in various ways, which is expected to affect the final behavior of the pesticide, with consequences on the efficacy [7]. Some of the first pesticide formulations developed in the agricultural industry (like granules, aqueous solutions, dusts, powders, and mineral oil in water emulsions) was based on simple technologies. However, since the 1980s, the pesticide industry has made great strides for the development of new formulations, focusing in particular on the search for greater chemical stability, optimization of biological activity, differentiation, and greater safety in use [8]. In addition, the search for decreasing the dose required per hectare to reduce the amounts of pesticides distributed in the environment has focused on the development of new formulations [9, 10]. The main factors that determine the design of a formulation are the solubility characteristics of the active ingredient (AI), cost of manufacture, and the intended use, so interdisciplinary sciences are required in each new formulation development [11].

The efficacy of agrochemicals as crop protection agents is generally a function of the intrinsic properties of the active ingredients, such as their toxicity, plant movement, penetration capacity, and mechanism of action [12] but also can be influenced by the formulation and the mode of application of the commercial product and the participation of surfactants and adjuvants among other parameters [13]. Formulation is a key tool because different formulations can promote stability to photochemical degradation, or decrease the amount of active ingredient necessary to achieve pest control [14]. Different works propose that a formulation can improve handling safety and can play a crucial role in the duration of delivery of the active ingredient [15, 16]. The formulation may also be a key point in avoiding phytotoxicity [17] or incompatibility on mixes with other agrochemicals [18].

The production of fruit in Chile corresponds to an industry focused on the export of fresh fruit [19], so it is subject to different phytosanitary requirements [20]. Within them, pest management is a relevant item, where the main management is carried out based on the chemical synthesis insecticides [21]. Due to the high rate of use of these products in developing countries like Chile [22], the chemical industry has found an attractive market, generating a wide range of insecticides, with several formulations of the same active ingredient. The above occurs, for example, with neonicotinoid insecticides acetamiprid and imidacloprid [23]; with the organophosphate insecticides diazinon [24], chlorpyrifos [25]; and with the pyrethroid insecticide lambda-cyhalothrin [26], among others. All these insecticides are commonly used in apple orchards in Chile.

Acetamiprid and imidacloprid are widely used to control obscure mealybug (*Pseudococcus viburni*) (Hemiptera: Pseudococcidae), San Jose scale (*Diaspidiotus perniciosus*) (Hemiptera: Diaspididae), wooly apple aphid (*Eriosoma lanigerum*) (Hemiptera: Aphididae) [27] and also, in the case of acetamiprid, is used to control codling moth (*Cydia pomonella*) (Lepidoptera: Tortricidae) on apple orchards [28, 29]. Diazinon is mainly used in the control of wooly apple aphid [30] and mealybugs [31] until the first stages of fruit development, while l-cyhalothrin is used up to the preharvest period to control codling moth [32]. The use of these pesticides is suggested based on their control objective, respecting a preharvest interval estimated to comply with the maximum residue limits. These intervals are currently estimated for the active ingredient independent of the formulation used [33], even when different works propose that formulations can affect dissipation of residues [34] and residue amount [35]. Likewise, formulation can affect efficacy of the application, generating a direct impact on the number of applications required to achieve adequate control [36, 37]. Also, formulation type might impact on the proportionality of residues, especially when changes in rate (kg active ingredient ha^{-1}) are accomplished by changing the spray concentration, because depending on the type of formulation, increasing spray rate will also increase surfactant and other adjuvant concentrations in the spray solution which can help the crop to retain for a longer period the residue [38]. About the influence on the efficacy, for example, a comparison performed between two formulations of imidacloprid and carbofuran found an increase in the control period of aphid and leafhopper in potato, when using encapsulated formulations of those insecticides compared with commercial formulations WP and G, respectively, but not in the dissipation of its residues [39].

The aim of this chapter is to evaluate effectiveness and residuality of two commercial formulations of the diazinon insecticide; two commercial formulations of acetamiprid, two commercial formulations of active ingredient lambda-cyhalothrin, and finally, three commercial formulations of imidacloprid insecticide. For diazinon and acetamiprid formulations, the comparison parameters used correspond to the efficacy in the simultaneous control of three primary importance pests in apple trees: *C. pomonella*, *D. perniciosus*, and *P. viburni*; for l-cyhalothrin formulations, efficacy against *C. pomonella* was compared; and for imidacloprid formulations, differences in control *P. viburni* were stablished.

2. Methodology

2.1. Insecticides

Assays were conducted using commercial formulations of insecticides. Then, diazinon 50% p/v emulsion in water (EW) emulsion (Diazol® 50 EW; Adama Makhteshim Ltd.) and diazinon 40% p/p wettable powder (WP) (Diazinon® 40 WP; Anasac Chile S.A.) were compared. Also was performed the comparison between acetamiprid 70% wettable powder (Hurricane® 70 WP; Anasac Chile S.A.) and 20% soluble powder (Mospilan® 20 SP; Nippon Soda Co., Ltd.). For lambda-cyhalothrin, 5% p/v microcapsule suspension (Karate Zeon® 050 CS; Syngenta S.A.) and 5% emulsifiable concentrate (EC) (Zero® 5 EC; Anasac Chile S.A.) formulations were compared, and finally, for imidacloprid insecticide, 20% p/p soluble liquid (SL) (Confidor® Forte 200 SL; Bayer CropScience AG), 35% p/v suspension concentrate (Confidor® 350 SC; Bayer CropScience AG), and 70% p/p wettable powder (Punto® 70 WP; Anasac Chile S.A.) were used.

2.2. Efficacy evaluations

During the spring of 2016, an apple orchard, cultivar *Royal Gala* located in the main pome fruit-growing area of Chile (34°46′45.9″S 71°02′50.0″W), was selected for this study. This orchard was naturally infested with the San Jose scale (*D. perniciosus*) and obscure mealybug (*P. viburni*). Prespraying evaluation was performed, determining that the appropriate statistical design was completely randomized with four replicates (each one with 50 plants, equivalents to 0.125 hectares).

The climatic conditions at the study period were as follows: average air temperature of 18.5°C (8.8–28.1°C) and relative humidity of 65.4% (33.2–97.6%). The first 22 days were free of precipitation, and then a total of 4 mm were recorded between days 23 and 25 post application. The phenological status at the beginning and the end of the study was 16 and 25 mm of diameter of fruits, respectively.

For *C. pomonella* evaluations, artificial infestations with neonate larvae (L_1) were performed on laboratory over 100 uninfested fruits collected per experimental unit. Neonate larvae were obtained from previous breeding in the laboratory, with insects coming from orchards not previously treated with insecticides. The fruits were collected from the experimental units at 3, 7, 10, 14, 21, and 25 days after application (DAA); collecting them from the pedicel to avoid the excessive manipulation of the residue of insecticides or removal. One larva was used per fruit, and mortality was recorded under microscope at 24 h post each infestation. Between infestation and evaluation, the fruits were maintained in breeding chamber at light conditions: darkness 16: 8 h, with 16 ± 2°C.

For *D. perniciosus* and *P. viburni* evaluations, the number of infested fruits and the number of live scales and live mealybugs were counted under microscope on 100 fruits collected per experimental unit, reaping 2 apples randomly per tree from each repetition at 3, 7, 10, 14, 21, and 25 DAA. In all cases to score insects as dead, failure of the insect to respond when probed with a dissecting needle, shriveling, and color variation was considered.

Mortality of codling moth larvae percentage was calculated for each insecticide and corrected using the Abbott's formula [40]. The data of efficacy on San Jose scale and obscure mealybug obtained from the experiment described above separately by active ingredient were subjected to analysis of variance (ANOVA) by taking appropriate transformations. Mean comparisons in significant ANOVAs were performed with a Tukey's test ($p \geq 0.05$). Statistical analyses were conducted using the software Minitab®16.1.0 (Minitab Inc.).

2.3. Treatments

A control treatment without insecticide applications was considered. In order to represent the use of insecticides under equal conditions, a single dose was used per active ingredient

(A.I.), given a total of 10 treatments and control included. Then, two treatments contained 50 g of diazinon/100 l of water; two treatments contained 8.4 g of acetamiprid/100 l; two treatments contained 1 cc of lambda-cyhalothrin/100 l; and three treatments contained 21 g of imidacloprid/100 l.

All applications were performed just once on the season, on November 2, with a conventional hydraulic sprayer (Line Ecofrut 2000, Parada S.A) dosing each treatment for 2000 l of water per hectare. Between treatments were left at least 30 m free of evaluations to avoid interference in the measurements.

2.4. Residue estimation of insecticides

Four apple samples (4 kg per experimental unit) for determination of each insecticide residues were taken at 1–25 DAA from all treatments [41]. Apple samples from each replicate of each treatment were chopped into small pieces and mixed, and subsample (100 g) was used for extraction.

Determination of acetamiprid and imidacloprid residues was done using P-002 Luke, method based on gas chromatography with mass detector (GC-MS) and high-performance liquid chromatography (HPLC) with triple quadrupole detector (MS/MS) [42, 43]. Determination of diazinon and l-cyhalothrin residues was done using gas chromatography (GC) with triple quadrupole detector (MS/MS) [44]. Finally, the data obtained on the initial and final deposits of different formulations of each active ingredient were subjected to ANOVA. For imidacloprid, mean comparisons were performed using ANOVA and Tukey's test ($p \geq 0.05$).

3. Results and discussion

3.1. Efficacy of diazinon formulations

Major knockdown effect and longer residual period to control *D. perniciosus* (**Figure 1**, **Table 1**) and *P. viburni* (**Figure 2**, **Table 2**) was achieved by using emulsion in water (EW) than wettable powder (WP). In addition, higher levels of mortality of both pests were achieved with the use of EW formulation. One work performed with diazinon against the attack of San Jose scale crawlers showed that diazinon provided 12–13 days of protection [45], which can be considered similar to results obtained on this chapter for WP formulation, but apparently it is underestimated for EW formulation. On both parameters (mean of infested fruits by scales or mealybugs and mean of living scales or mealybugs on fruits), EW seems to be effective even until the last evaluation carried out at 25 DAA.

On the other hand, about *C. pomonella* control (**Figure 3**), both formulations showed and optimal and similar control until 10 DAA, and then, better results—but not optimal—were obtained with EW formulation. One work conducted in 1965 proposed that for diazinon, optimal insecticide activity against *C. pomonella* would have an approximate duration of 6 days [46]; in the present work, demanding for 90% minimum of larvae mortality, both formulations deliver 10 days of control. On 14 DAA evaluations, EW formulation showed a mortality level close to 80%, which is considered insufficient from the economic point of view for the

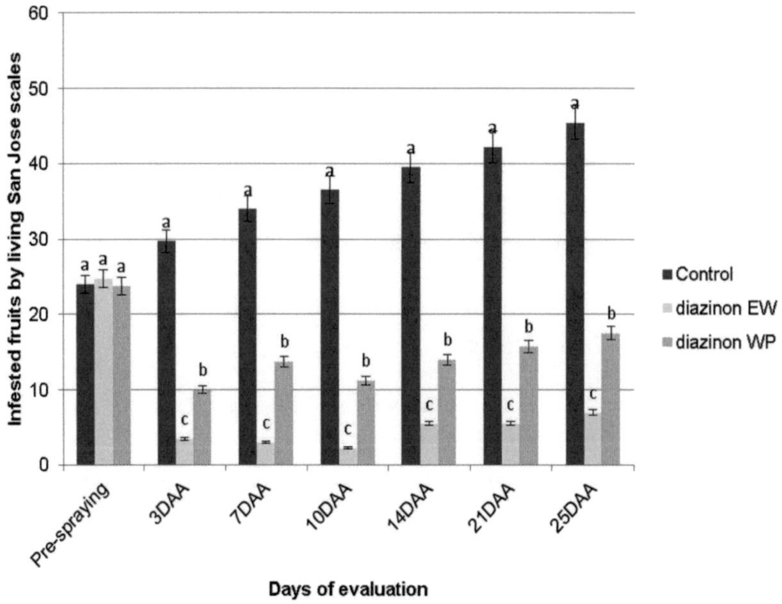

Figure 1. Mean of infested fruit by living San Jose scale according to the treatment of diazinon formulations. Values followed by different letters indicate significant differences with $p < 0.05$ (ANOVA, Tukey's test).

farmer. Therefore, although both formulations show significant differences in the control of *C. pomonella*, commercially (demanding a mortality of at least 85%), both only control efficiently for up to 10 days.

3.2. Residues of diazinon formulations

Emulsion in water generates higher initial and final diazinon residues than wettable powder (**Figure 4**). These results are probably due to differences between formulations that affect

Mean of living scales (*D. perniciosus*) on fruits according to diazinon formulation by evaluation moment (DAA)

Treatment	Prespraying	3DAA	7DAA	10DAA	14DAA	21DAA	25DAA
Control	31.00a	36.00a	39.25a	40.00a	43.75a	47.75a	51.75a
Emulsion in water	29.25a	7.25c	4.75c	3.25c	6.75c	7.75c	8.25c
Wettable powder	27.50a	11.75b	14.75b	12.75b	12.00b	17.50b	19.25b
F	**1.08**	**200.45**	**78.63**	**72.98**	**172.79**	**155.83**	**137.18**
***p* value**	**0.379**	**<0.001**	**<0.001**	**<0.001**	**<0.001**	**<0.001**	**<0.001**

Means followed by different letters indicate significant differences with $p < 0.05$ (ANOVA, Tukey's test).

Table 1. Mean of living San Jose scale on fruits according to the treatment of diazinon formulations.

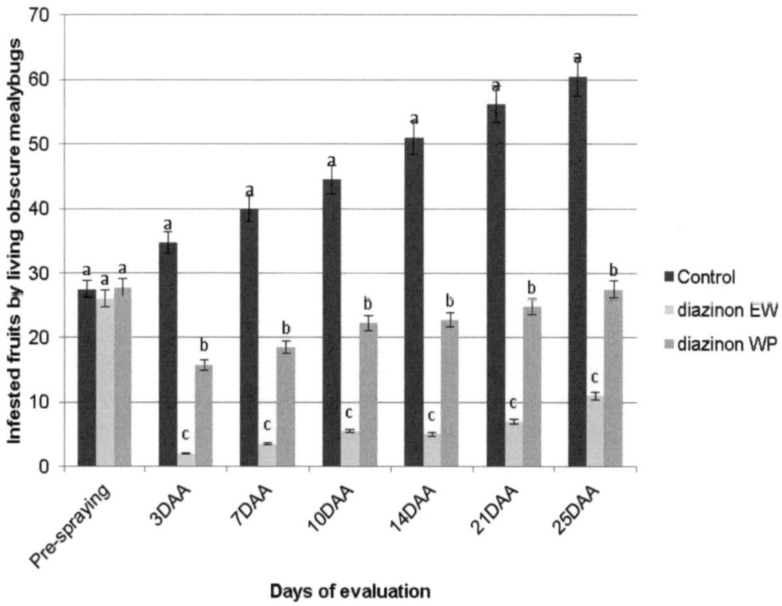

Figure 2. Mean of infested fruit by living obscure mealybugs according to the treatment of diazinon formulations. Values followed by different letters indicate significant differences with $p < 0.05$ (ANOVA, Tukey's test).

the surface tension of the solution and hence droplet formation and deposition of diazinon residue. Diazinon is a water-soluble insecticide with high affinity for lipids [47]; then, solvent and emulsifier used on emulsion in water formulations can play a crucial role on the deposition pattern. Emulsion in water dissolved in water forms an emulsion, which does not need constant agitation to maintain it; instead, the wettable powder formulation forms a suspension, which requires constant agitation to keep its fine particles suspended in the water. These differences in the physical-chemical behavior of both formulations were reflected in differences in their initial deposition and persistence of its residues, but this does not seem to be a constant to all types of formulations and insecticides. One work comparing residue levels generated by three formulations of chlorpyrifos (emulsifiable concentrate (EC), wettable granules (WG),

Mean of living obscure mealybugs on fruits according to diazinon formulation by evaluation moment (DAA)							
Treatment	**Prespraying**	**3DAA**	**7DAA**	**10DAA**	**14DAA**	**21DAA**	**25DAA**
Control	33.50a	42.75a	48.00a	55.00a	57.00a	61.50a	62.25a
Emulsion in water	31.25a	3.00c	6.00c	5.75c	5.50c	7.25c	12.50c
Wettable powder	32.75a	18.75b	21.25b	24.75b	24.50b	26.50b	32.25b
F	**0.55**	**80.36**	**175.45**	**279.37**	**129.88**	**178.28**	**142.53**
p **value**	**0.597**	**<0.001**	**<0.001**	**<0.001**	**<0.001**	**<0.001**	**<0.001**

Means followed by different letters indicate significant differences with $p < 0.05$ (ANOVA, Tukey's test).

Table 2. Mean of living obscure mealybugs on fruits according to the treatment of diazinon formulations.

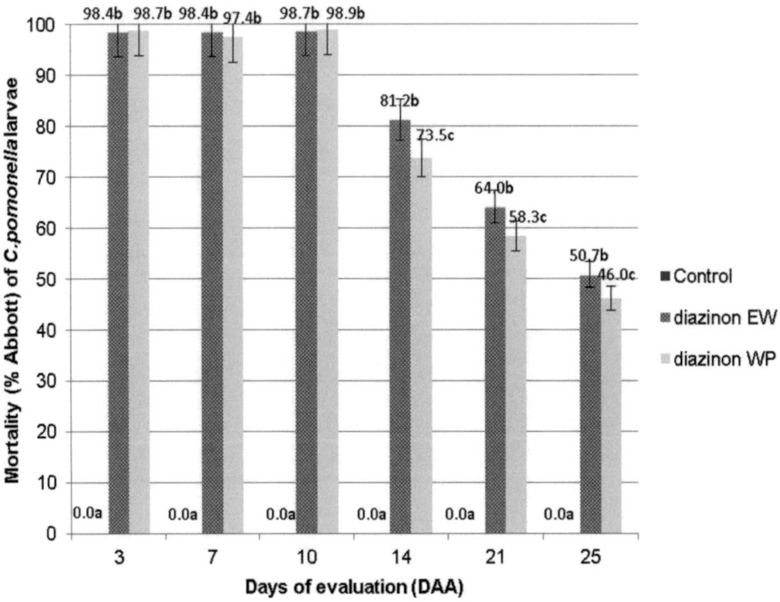

Figure 3. Mortality (% Abbott) of *C. pomonella* larvae according to the treatment of diazinon formulations. Values followed by different letters indicate significant differences with $p < 0.05$ (ANOVA, Tukey's test).

and microencapsulates (ME)) applied to oranges shows that the decline curve and the residue levels in fruits, leaves, and soil could change remarkably if the same active ingredient is used in different formulations [48]; on contrary, the study performed with fenitrothion applied to oranges and clementines with emulsifiable concentrate and microencapsulate formulations did not find differences on rate of decline residue of the active ingredient for both kinds of commercial formulations [49]. For the insecticide azadirachtin, EC formulations compared

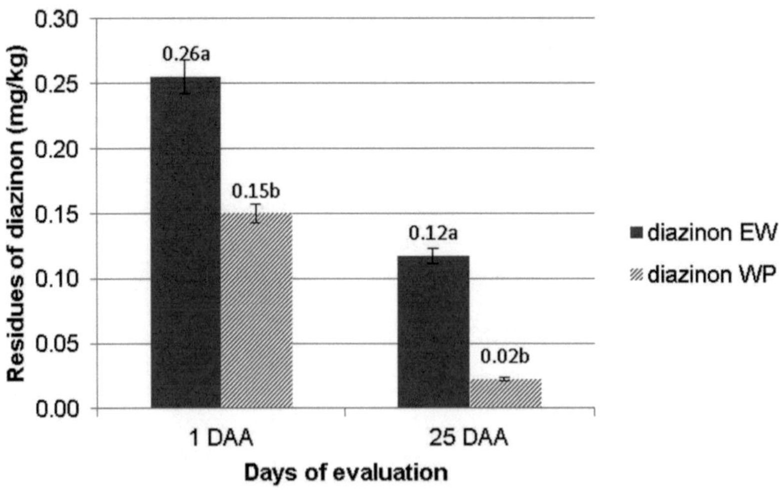

Figure 4. Initial and final residues of diazinon quantified for each formulation. Values followed by different letters indicate significant differences with $p < 0.05$ (ANOVA).

with WP formulation showed differences in the droplet-size spectra and deposit levels, attributed to the influence of additives present in different formulations [50].

3.3. Efficacy of acetamiprid formulations

Differences on efficacy only after 21 DAA were observed with acetamiprid formulations on codling moth (**Figure 5**). Major knockdown effects on San Jose scale (**Figure 6**; **Table 3**) and obscure mealybug (**Figure 7**; **Table 4**) were obtained using wettable powder (WP); nevertheless a longer protection period was obtained using soluble powder (SP). On *C. pomonella*, both treatments showed appropriate control until 21 DAA, and then, only soluble powder maintained a percentage of control with over 85% larvae mortality until 25 DAA.

Acetamiprid has shown good control activity against *C. pomonella* [51]. It is systemic and intended to control sucking insects like aphids, mealybugs [52], and San Jose scale [53]. In this study, differences on the behavior on control could be explained by differences on chemical-physical property between both formulations. Although they have many similarities, when mixed on water, WP generates a suspension, and when applied, formulation particles remain on the treated surface [54]; instead, SP generates a homogenous solution which is easily incorporated by the plant. In addition, between both formulations, the concentration of active ingredient and the proportion and type of coformulants vary. Then, it is possible that these differences are reflected in different rates of absorption by the plant and/or insects [55]. There is some consensus that biological performance of a pesticide is frequently affected by the choice of formulation type [56, 57], for example, a formulation which delivers the chemical in

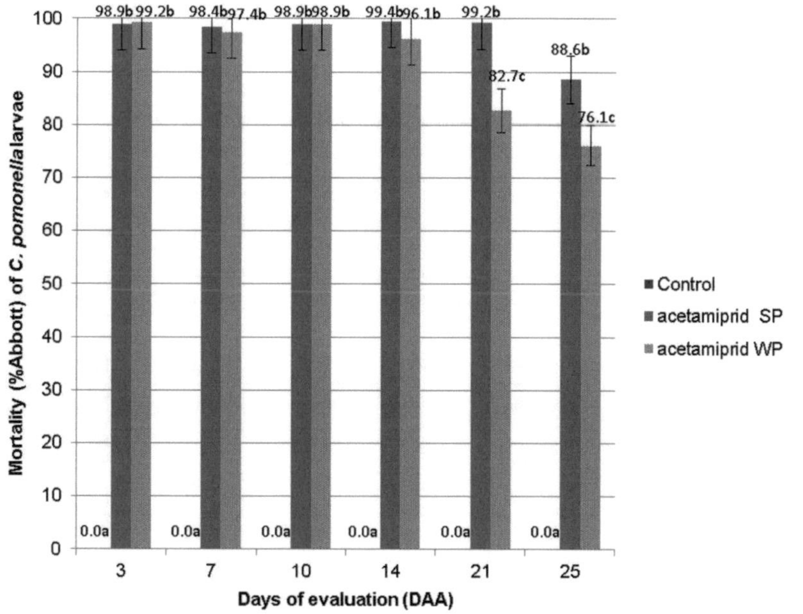

Figure 5. Mortality (% Abbott) of *C. pomonella* larvae according to the treatment of acetamiprid formulations. Values followed by different letters indicate significant differences with $p < 0.05$ (ANOVA, Tukey's test).

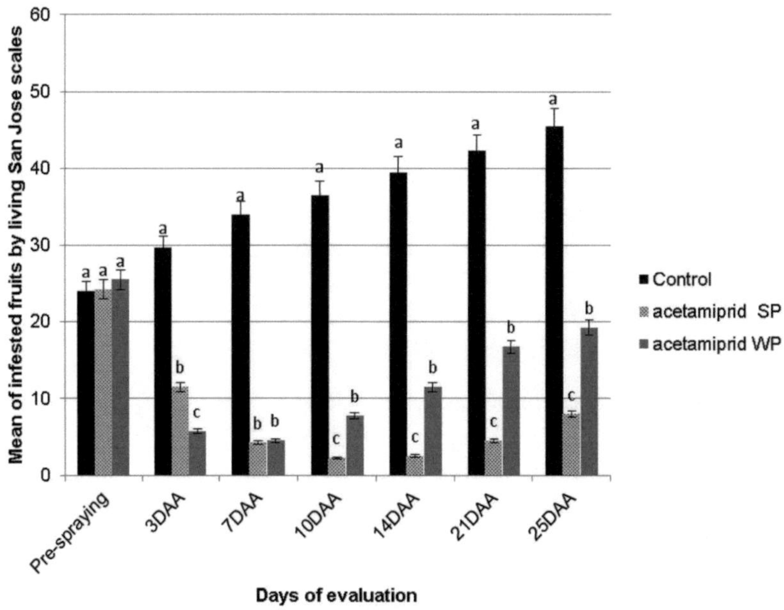

Figure 6. Mean of infested fruit by living San Jose scale according to the treatment of acetamiprid formulations. Values followed by different letters indicate significant differences with $p < 0.05$ (ANOVA, Tukey's test).

a solution, as in SL, EC, or SP formulations, is commonly considered more biologically active than WP or SC formulations, but also, it has a greater risk of being phytotoxic [58].

3.4. Residues of acetamiprid formulations

Wettable powder formulation generated higher initial deposition of acetamiprid than observed for the soluble powder formulation (**Figure 8**). These results contradict what was proposed for a comparison between decline curves of acetamiprid on apple (cv. Pink Lady) performed with SL and WP formulations. In that work, no significant differences were found on the initial and final depositions of acetamiprid between both formulations, applied on

Mean of living scales (*D. perniciosus*) on fruits according to acetamiprid formulation by evaluation moment (DAA)							
Treatment	**Prespraying**	**3DAA**	**7DAA**	**10DAA**	**14DAA**	**21DAA**	**25DAA**
Control	31.00a	36.00a	39.25a	40.00a	43.75a	47.75a	51.75a
Soluble powder	28.25a	11.50b	6.50c	2.25c	4.50c	6.00c	9.25c
Wettable powder	30.25a	6.00c	13.50b	8.25b	12.50b	24.25b	27.75b
F	**1.22**	**204.07**	**79.46**	**169.31**	**178.52**	**127.69**	**102.01**
***p* value**	**0.34**	**<0.01**	**<0.01**	**<0.01**	**<0.01**	**<0.01**	**<0.01**

Means followed by different letters indicate significant differences with $p < 0.05$ (ANOVA, Tukey's test).

Table 3. Mean of living San Jose scale on fruits according to the treatment of acetamiprid formulations.

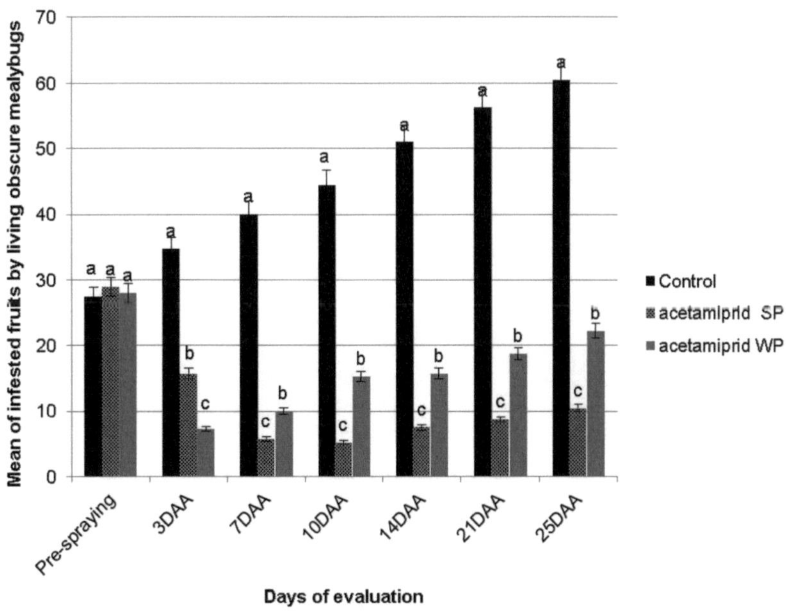

Figure 7. Mean of infested fruit by obscure mealybugs according to the treatment of acetamiprid formulations. Values followed by different letters indicate significant differences with $p < 0.05$ (ANOVA, Tukey's test).

three different apple growth stages [59]. Even when both SP and SL form solutions, these formulations differ on its coformulants and the concentration of active ingredient contained proportionally therein. Thus, even if in this work an equal amount of active ingredient per hectoliter per hectare has been dosed, the proportion and type of surfactants, carriers, or others will not be equivalent.

In general terms, results from the literature are not conclusive with respect to the effect of different formulations in residue deposits and behavior of pesticides, perhaps because of the difficulty of isolating other factors that also affect the degradation of residues such as species and varieties; use of adjuvants; types and concentration of coformulants; fruit growth;

Mean of living obscure mealybugs on fruits according to acetamiprid formulation by evaluation moment (DAA)							
Treatment	**Prespraying**	**3DAA**	**7DAA**	**10DAA**	**14DAA**	**21DAA**	**25DAA**
Control	33.50a	42.75a	48.00a	55.00a	57.00a	61.50a	62.25a
Soluble powder	29.00a	15.75b	7.00b	5.50c	7.75c	9.00c	10.75c
Wettable powder	28.00a	7.25c	11.25b	15.25b	17.00b	18.75b	22.25b
F	**2.60**	**79.16**	**282.59**	**282.08**	**168.09**	**234.43**	**238.61**
***p* value**	**0.13**	**<0.01**	**<0.01**	**<0.01**	**<0.01**	**<0.01**	**<0.01**

Means followed by different letters indicate significant differences with $p < 0.05$ (ANOVA, Tukey's test).

Table 4. Mean of living obscure mealybugs on fruits according to the treatment of acetamiprid formulations.

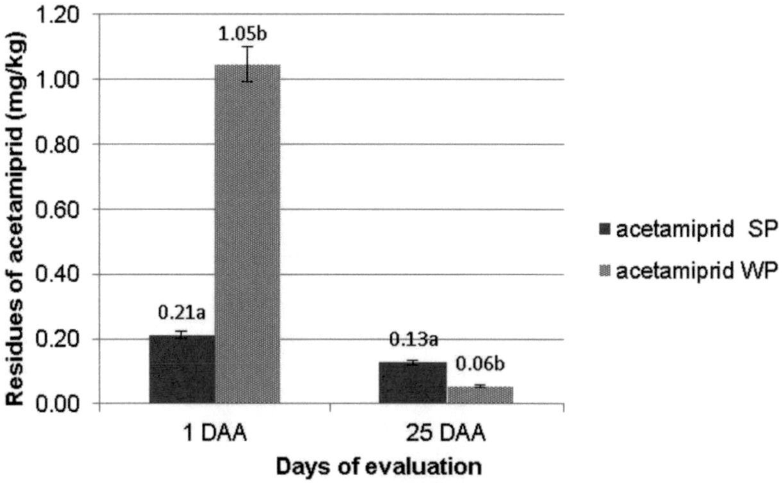

Figure 8. Initial and final residues of acetamiprid quantified for each formulation. Values followed by different letters indicate significant differences with $p < 0.05$ (ANOVA).

climatic conditions; spraying method; measuring technique; and pesticide physicochemical properties between others [60–62].

3.5. Efficacy of l-cyhalothrin formulations

Although both formulations showed insecticidal activity against *C. pomonella*, significant differences between both formulations were observed since the first evaluation (**Figure 9**). Zeon

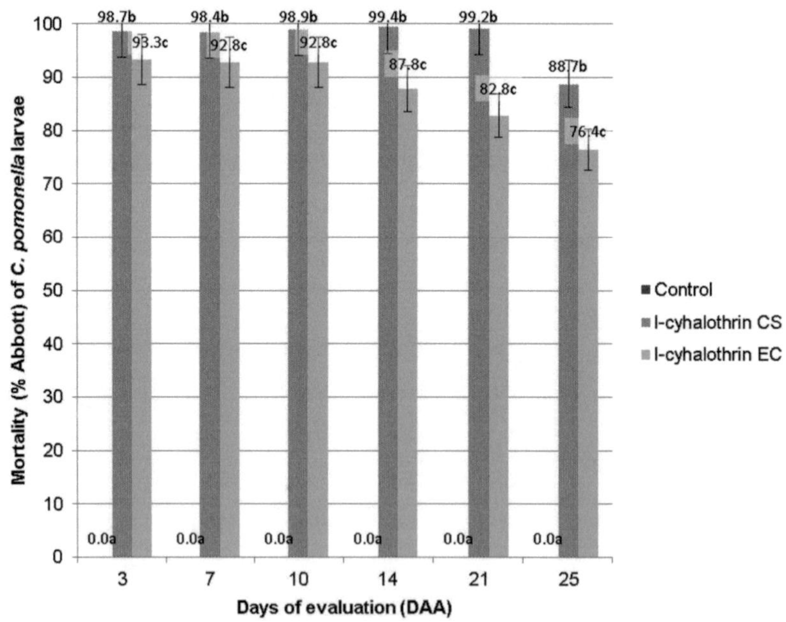

Figure 9. Mortality (% Abbott) of *C. pomonella* larvae according to the treatment of l-cyhalothrin formulations. Values followed by different letters indicate significant differences with $p < 0.05$ (ANOVA, Tukey's test).

(CS) formulation showed better efficacy, with the longer stable control period. Key difference between Zeon technology and EC formulation is that the first one encapsulates the active ingredient in small capsules with thin walls; instead, the second one comprises the active ingredient, a solvent, and emulsifiers. This enables for Zeon formulation quick "knockdown" of insects coupled with long-term persistence [63, 64], properties that were observed in this study. Even when they observed differences between efficacy parameters of both formulations, some authors propose that other parameters as application rate should be considered [65], which, in the present study, is a constant factor for the comparison of both formulations allowing us to conclude without other variables. One work conducted to evaluate efficacy against *Cydia molesta* founded that in the laboratory and in the field, the toxicity to *C. molesta* larvae of microencapsulated (CS) l-cyhalothrin was similar to that of the emulsifiable concentrate (EC) formulation. However, the same work proposed that different toxicity responses were obtained when evaluating its effect on the predator *Typhlodromus pyri* Scheuten, where CS formulation results significantly more toxic than EC formulation in pyrethroid-susceptible predator population [66].

3.6. Residues of l-cyhalothrin formulations

Apparently, the different release rates of the active ingredient are not significantly different at their start (1 DAA) and at 25 days from the application (**Figure 10**), possibly because the microencapsulated formulations do not necessarily vary the deposition of the active ingredient residue, but their biological availability and, as it was observed, its effectiveness.

3.7. Efficacy of imidacloprid formulations

Control of *P. viburni* showed by soluble liquid (SL) and suspension concentrate formulations was similar to wettable powder (WP) formulation until 10 DAA, reflected in both the average

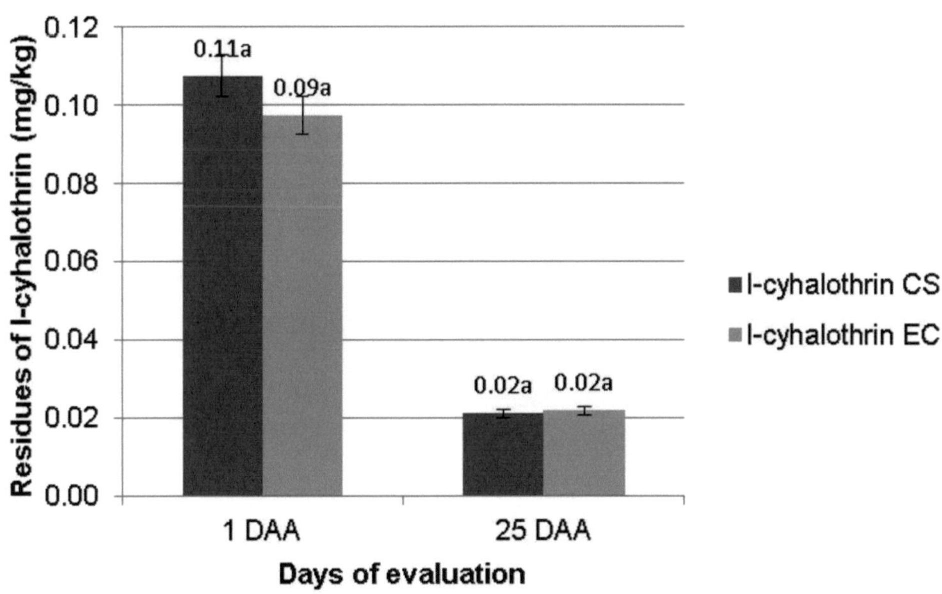

Figure 10. Initial and final residues of l-cyhalothrin quantified for each formulation. Values followed by different letters indicate significant differences with $p < 0.05$ (ANOVA).

number of fruits infested by the pest (**Figure 11**) and the mean of living mealybugs on fruits (**Table 5**); then, longer efficient period of control was obtained using SL formulation.

Imidacloprid is highly effective against mealybugs [67] and others hemipteran pest [68, 69]. It is available in different formulations (WP, SL, SC, OD, WG) registered generally according to their use intention (foliar sprays, seed treatments, and via soil application) [70]; but for the same target of control and way of use, more than one formulation can be available. The choice of one formulation or another may vary the metabolism and persistence behavior of imidacloprid [71, 72]. In our work, even though the three formulations differ in the characteristics of their coformulants, WP and SC formulations have in common that both form suspensions on water; instead, SL forms a solution. This difference could generate different responses in the mobility and translocation of the active ingredient, and therefore its availability to control the pest.

3.8. Residues of imidacloprid formulations

Both initial and final deposits of imidacloprid were higher in SL, compared with WP and SC (**Figure 12**). Even though this work presents only two points or moments of evaluation within a possible residue decline curve, we can infer that the degradation of imidacloprid residue was affected by the formulation, resulting more persistently the residue of imidacloprid generated by the formulation SL than that of the other formulations WP and SC.

On the other hand, the formulations WP and SC had similar behaviors with each other, which is coincident with what was observed in the construction of metalaxyl decline curves in grapes for WP and SC formulations of this pesticide [73].

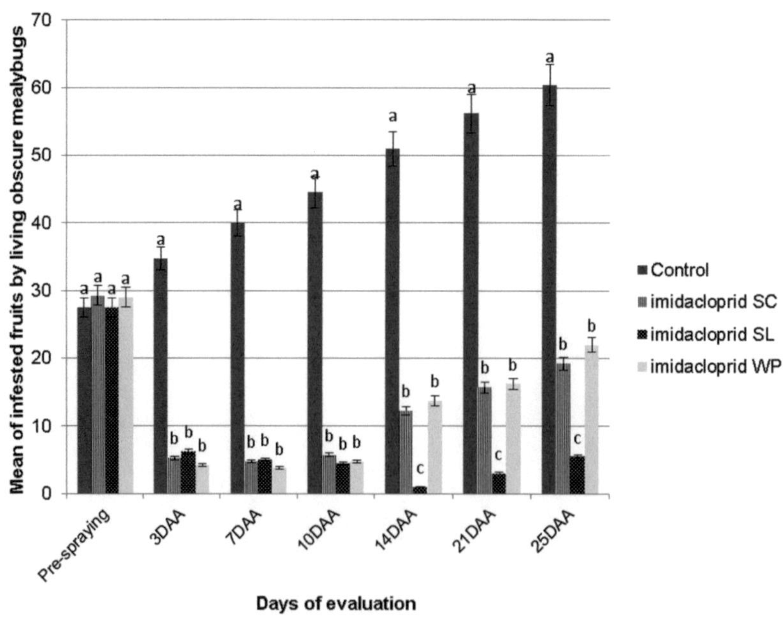

Figure 11. Mean of infested fruit by obscure mealybugs according to the treatment of imidacloprid formulations. Values followed by different letters indicate significant differences with $p < 0.05$ (ANOVA, Tukey's test).

Mean of living obscure mealybugs on fruits according to imidacloprid formulation by evaluation moment (DAA)							
Treatment	**Prespraying**	**3DAA**	**7DAA**	**10DAA**	**14DAA**	**21DAA**	**25DAA**
Control	33.50a	42.75a	48.00a	55.00a	57.00a	61.50a	62.25a
Suspension concentrate	36.50a	5.25b	4.75b	6.25b	12.75b	16.00b	19.75b
Soluble liquid	36.00a	6.25b	5.00b	5.00b	1.75c	3.50c	6.75c
Wettable powder	35.75a	4.25b	3.75b	4.75b	14.5b	17.00b	24.50b
F	**0.51**	**10.12**	**29.31**	**33.09**	**20.82**	**23.49**	**17.13**
***p* Value**	**0.68**	**<0.001**	**<0.001**	**<0.001**	**<0.001**	**<0.001**	**<0.001**

Means followed by different letters indicate significant differences with $p < 0.05$ (ANOVA, Tukey's test).

Table 5. Mean of living obscure mealybugs on fruits according to the treatment of imidacloprid formulations.

In all the cases studied on this research (except both formulations of l-cyhalothrin), concentration of active ingredient contained in 1 l or kg of each commercial product varied with their respective comparison. Thus, even though all treatments were performed to look for the same dose of active ingredient per hectare, there was always a variation in the content of adjuvants, surfactants, inerts, or other contents according to the formulation. In the case of l-cyhalothrin, the variations passed through the type of coformulants in addition to the differentiating characteristics of the rate of release of the active ingredient, but not in the concentration of active ingredient. For Food and Agriculture Organization (FAO), two pesticides may not be classified as equivalent even when they have the same concentration of active ingredient, as their similarity depends on the type of formulations. Nor does it apply nominally similar products from other manufacturers or at all those where the active ingredient is produced

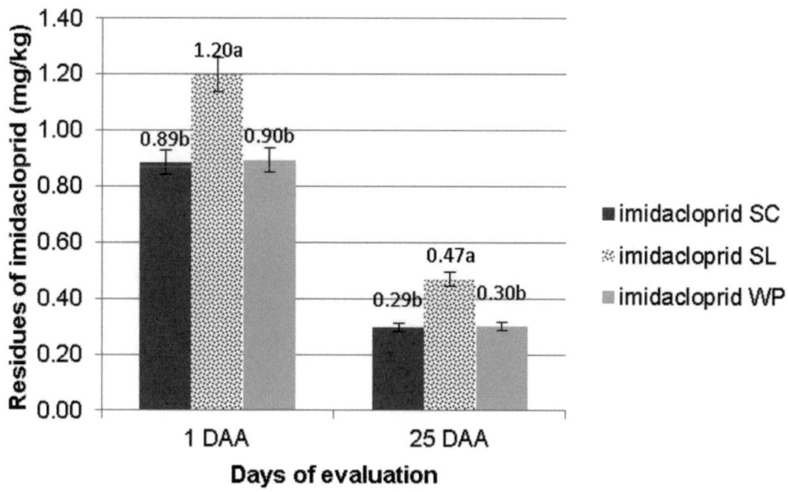

Figure 12. Initial and final residues of imidacloprid quantified for each formulation. Values followed by different letters indicate significant differences with $p < 0.05$ (ANOVA).

by other synthesis methods [74]; inferring that, it is expected that there are certain variations in its practical behavior, whether in its pattern of residues, biological behavior, or efficacy. Accordingly, in the pesticide registry and its establishment of maximum residue limits, it is suggested that information is submitted for the formulation to be registered, and, if a new formulation is to be introduced, it is suggested to carry out collaborative trials (even between different manufacturers) that previously demonstrate that the variations made in the new formulation will not change the pattern of residues [41]. The information generated in these studies is not public.

While the chemical industry posits that some formulations do not imply significant variations between them [74], there is a lot of confidential information that is not known to the end user, which prevents easily discriminating when variations are expected or when not. For the same reason, and because the effectiveness or protection periods may be severely affected by variations between formulations, further comparative inquiries are required to discriminate between product profiles for pest control.

4. Conclusions

Pesticide formulations would have a significant impact on the biological effect for the studied pesticides. Efficacy, knockdown effect and period of effectiveness protection can be affected by formulation of an insecticide. All these parameters are determinant in the design of an application program of insecticides, but in most cases are unknown by the user, the farmer, generating consequently that a greater number of applications are realized. This increases the impact on the environment and likely the costs associated with pest management.

On the other hand, when insecticide residue can be affected by formulation (both in the initial deposit and in its persistence), other parameters inherent to the active ingredient and the capacity of detection and quantification of the measurement technique may be mediating the real importance that the formulation may have on the behavior of the residue. In this chapter, the formulation was determinant in the residual pattern of acetamiprid, imidacloprid, and diazinon, but not for l-cyhalothrin. Future complementary works may focus on considering other interaction between variables such as relevance of the type and concentration of coformulants; fruit growth stage; climatic conditions and spraying method; and fruit or vegetable species, since all of them dynamically seem to affect in some degree the behavior of the insecticide residue generated. This in turn suggests that in some cases the estimation of preharvest intervals calculated for certain active ingredients may be affected by the formulation used in the baseline studies, generating the risk of an underestimation of that interval.

Therefore, when formulations of the same active ingredient are widely available, it is desirable to have independent declination curves of insecticide residue for at least those which differ drastically in type of coformulants or adjuvants and its concentration of active ingredient, considering that there is a risk of significant variations in the behavior of the residue.

http://dx.doi.org/10.5772/intechopen.72340

Author details

Karina Buzzetti

Address all correspondence to: karinabuzzetti@agridev.cl

Consultant Agri Development Ltd., Santiago, Chile

References

[1] Hall ER. Spray deposits: Opportunities for improved efficiency of utilization v/a quality. Quantity and Formulation. Phytoparasitica. 1997;**25**:39S-52S

[2] Stephenson GR, Ferris IG, Holland PT, Nordberg M. Glossary of terms relating to pesticides. (IUPAC recommendations 2006). Pure and Applied Chemistry. 2006;**78**:2075-2154

[3] Aktar MW, Sengupta D, Chowdhury A. Impact of pesticides use in agriculture: Their benefits and hazards. Interdisciplinary Toxicology. 2009;**2**:1-12

[4] Coelho S. European pesticide rules promote resistance, researchers warn. Science. 2009;**323**:450

[5] Lorman A. Chapter 3: Encouraging improved pesticide formulation: Data protection for inert ingredients. In: Foy CL, Pritchard DW, editors. Pesticide Formulation and Adjuvant Technology. CRC Press; 1996. pp. 25-32

[6] FAO. International Code of Conduct on the Distribution and Use of Pesticides. Rome: Food and Agricultural Organisation; 2005. 36 p

[7] van Valkenburg W, editor. Pesticide formulations. New York: Marcel Dekker; 1973. 481 p

[8] Carmine S. Pesticide Formulation and Delivery Systems: 33rd Volume, "Sustainability: Contributions from Formulation Technology". ASTM International; 2014. 185 p

[9] Knowles A. Recent developments of safer formulations of agrochemicals. The Environmentalist. 2008;**28**:35-44

[10] Lagaly G. Pesticide–clay interactions and formulations. Applied Clay Science. 2001;**18**: 205-209

[11] Kah M, Beulke S, Tiede K, Hofmann T. Nanopesticides: State of knowledge, environmental fate, and exposure modeling. Critical Reviews in Environmental Science and Technology. 2013;**43**:1823-1867

[12] Stenersen J. Chemical Pesticides Mode of Action and Toxicology. CRC Press; 2004. 277 p

[13] Wang CJ, Liu ZQ. Foliar uptake of pesticides—Present status and future challenge. Pesticide Biochemistry and Physiology. 2007;**87**:1-8

[14] Marrs GJ, Middleton MR. The formulation of pesticides for convenience and safety. Outlook on Agriculture. 1973;**7**:231-235

[15] Leonas KK. Effect of pesticide formulation on transmission: A comparison of three formulations. Bulletin of Environmental Contamination and Toxicology. 1991;**46**:697-704

[16] Tsuji K. Microencapsulation of pesticides and their improved handling safety. Journal of Microencapsulation. 2001;**18**:137-147

[17] Diehl KE, Stoller EW. Effect of simulated rainfall, insecticide formulation, and insecticide application method on the interaction between nicosulfuron and terbufos in corn (*Zea mays*). Weed Technology. 1995;**9**:80-85

[18] Lancaster SH, Jordan DL, Johnson PD. Influence of graminicide formulation on compatibility with other pesticides. Weed Technology. 2008;**22**:580-583

[19] Gwynne RN. Globalisation, commodity chains and fruit exporting regions in Chile. Tijdschrift voor Economische en Sociale Geografie. 1999;**90**:211-225

[20] Melo O, Engler A, Nahuehual L, Cofre G, Barrena J. Do sanitary, phytosanitary, and quality-related standards affect international trade? Evidence from Chilean fruit exports. World Development. 2014;**54**:350-359

[21] Buzzetti K, Chorbadjian RA, Nauen R. Resistance Management for San Jose Scale (Hemiptera: Diaspididae). Journal of Economic Entomology. 2015;**108**:2743-2752

[22] Ecobichon DJ. Pesticide use in developing countries. Toxicology. 2001;**160**:27-33

[23] Elbert A, Haas M, Springer B, Thielert W, Nauen R. Applied aspects of neonicotinoid uses in crop protection. Pest Management Science. 2008;**64**:1099-1105

[24] Narkiewicz-Jodko J, Rogowska M. The effectiveness of new formulations of diazinon: Basudin 600 EW and Diazol 50 EW in the control of large white butterfly (*Pieris brassicae* L.). Vegetable crops. Research Bulletin. 2000;**52**:69-73

[25] Solomon KR, Williams WM, Mackay D, Purdy J, Giddings JM, Giesy JP. Properties and uses of chlorpyrifos in the United States. In: Giesy JP, Solomon KR, editors. Ecological Risk Assessment for Chlorpyrifos in Terrestrial and Aquatic Systems in the United States. Springer International Publishing; 2014. pp. 13-34

[26] He LM, Troiano J, Wang A, Goh K. Environmental chemistry, ecotoxicity, and fate of lambda-cyhalothrin. Reviews of Environmental Contamination and Toxicology. 2008;**195**:71-91

[27] Kodandaram MH, Rai AB, Haldar J. Novel insecticides for management of insect pest in vegetable crops: A review. Journal of Vegetation Science. 2010;**37**:109-123

[28] Knight AL. Cross-resistance between azinphos-methyl and acetamiprid in populations of codling moth, *Cydia pomonella* (L.)(Lepidoptera: Tortricidae), from Washington State. Pest Management Science. 2010;**66**:865-874

[29] Brunner JF, Beers EH, Dunley JE, Doerr M, Granger K. Role of Neonicotinyl insecticides in Washington Apple Integrated Pest Management. Part I. Control of Lepidopteran Pests. Journal of Insect Science. 2005;**5**:14

[30] Carnegie AJM. Woolly aphid of apple, *Eriosoma lanigerum* (Hsm.), and its control in southern Rhodesia. Bulletin of Entomological Research. 1963;**53**:609-619

[31] Agnello AM, Spangler SM, Reissig WH, Lawson DS, Weires RW. Seasonal development and management strategies for Cornstock Mealybug (Homoptera: Pseudococcidae) in New York Pear Orchards. Journal of Economic Entomology. 1992;**85**:212-225

[32] Mota-Sanchez D, Wise JC, Poppen RV, Gut LJ, Hollingworth RM. Resistance of codling moth, *Cydia pomonella* (L.)(Lepidoptera: Tortricidae), larvae in Michigan to insecticides with different modes of action and the impact on field residual activity. Pest Management Science. 2008;**64**:881-890

[33] González RH. Degradación de residuos de plaguicidas en huertos frutales en Chile. Serie de Ciencias Agronómicas N° 4. Facultad de Ciencias Agronómicas, Universidad de Chile, Santiago, Chile: Ograma; 2002. 163 p

[34] Cabras P, Meloni M, Gennari M, Cabitza F, Ubeddu M. Pesticide residues in lettuce. 2. Influence of formulations. Journal of Agricultural and Food Chemistry. 1989;**37**:1405-1407

[35] Angioni A, Dedola F, Garau A, Sarais G, Cabras P, Caboni P. Chlorpyrifos residues levels in fruits and vegetables after field treatment. Journal of Environmental Science and Health, Part B. 2011;**46**:544-549

[36] Korunic Z. Review diatomaceous earths, a group of natural insecticides. Journal of Stored Products Research. 1998;**34**:87-97

[37] Arthur FH. Evaluation of a new insecticide formulation (F2) as a protectant of stored wheat, maize, and rice. Journal of Stored Products Research. 2004;**40**:317-330

[38] MacLachlan D, Hamilton DA. Review of the different application rates on pesticide residue levels in supervised residue trials. Pest Management Science. 2011;**67**:609-615

[39] Kumar J, Shakil N, Khan M, Malik K, Walia S. Development of controlled release formulations of carbofuran and imidacloprid and their bioefficacy evaluation against aphid, Aphis gossypii and leafhopper, Amrasca biguttula biguttula Ishida on potato crop. Journal of Environmental Science and Health, Part B. 2011;**46**:678-682

[40] Abbott WSA. Method of computing the effectiveness of an insecticide. Journal of Economic Entomology. 1925;**18**:265-267

[41] FAO. Guidelines on Producing Pesticide Residues Data from Supervised Trials, Rome. 1990

[42] Luke MA, Doose GMA. Modification of the Luke multiresidue procedure for low moisture, non-fatty products. Bulletin of Environmental Contamination and Toxicology. 1983;**30**:110-116

[43] Luke M, Froberg JE, Masumoto HE. Extraction and clean-up of organochlorine, organophosphate, organonitrogen and hydrocarbon pesticides in produce for determination by

gas–liquid chromatography. Journal of the Association of Official Analytical Chemists. 1975;**58**:1020-1026

[44] Alder L, Greulich K, Kempe G, Vieth B. Residue analysis of 500 high priority pesticides: Better by GC–MS or LC–MS/MS? Mass Spectrometry Reviews. 2006;**25**:838-865

[45] Thakur A, Hameed S. Toxicity and persistence of some organophosphorous insecticides against San Jose scale Quadraspidiotus perniciosus Comstock. Journal of the Indian Institute of Science. 1981;**63**:121-129

[46] Gratwick M, Sillibourne JM, Tew RP. The toxicity of insecticides to larvae of the codling moth, Cydia pomonella (L.). I.—Intrinsic toxicity and persistence. Bulletin of Entomological Research. 1965;**56**:367-376

[47] O'Neil MJ, editor. The Merck Index - An Encyclopedia of Chemicals, Drugs, and Biologicals. Whitehouse Station, NJ: Merck and Co., Inc; 2006. p. 509

[48] Montemurro N, Grieco F, Lacertosa G, Visconti A. Chlorpyrifos decline curves and residue levels from different commercial formulations applied to oranges. Journal of Agricultural and Food Chemistry. 2002;**50**:5975-5980

[49] Montemurro N, Grieco F, Lacertosa G, Visconti A. Persistence of fenitrothion in oranges and clementines after treatment with emulsifiable concentrate and microencapsulate formulations. Food Additives and Contaminants. 2005;**22**:39-47

[50] Sundaram KMS. Effect of additives in the neem formulations on deposition, volatilization and persistence of azadirachtin in spruce foliage. Journal of Environmental Science & Health Part B. 1997;**32**:523-544

[51] Magalhaes LC, Walgenbach JF. Life stage toxicity and residual activity of insecticides to codling moth and oriental fruit moth (Lepidoptera: Tortricidae). Journal of Economic Entomology. 2011;**104**:1950-1959

[52] Franco JC, Suma P, Da Silva EB, Blumberg D, Mendel Z. Management strategies of mealybug pests of citrus in Mediterranean countries. Phytoparasitica. 2004;**32**:507-522

[53] Frăsin LBN, Sumedrea M. Research on the efficacy of some insecticides in San Jose scale-Quadraspidiotus perniciosus Comst. Control in Maracineni-arges fruit growing area. Annals: Food Science and Technology. 2013;**14**:405-409

[54] Couch TL, Ignoffo CM. Formulation of insect pathogens. In: Burges HD, editor. Microbial Control of Pests and Plant Diseases. New York: Academic Press; 1981. pp. 621-634

[55] Becker N, Petrić D, Boase C, Lane J, Zgomba M, Dahl C, Kaiser A. Chemical control. In: Mosquitoes and their Control. Springer US; 2003. pp. 377-405

[56] Kumar J, Singh G, Palta RK, Walia S, Parsad R, Parmar BS. Field appraisal of controlled release formulations of carbofuran against the rice leaffolder, cnaphalocrocis medinalis (guenee). Indian Journal of Agricultural Science. 2006;**76**:732-735

[57] Kumar J, Shakil NA, Chandra S, Walia S, Shukla L, Parmar BS. Field appraisal of controlled release formulations of cartap hydrochloride against the rice leaf-folder, Cnaphalocrocis medinalis. Indian Journal of Agricultural Science. 2010;**80**:56-59

[58] Seaman D. Trends in the formulation of pesticides—An overview. Pest Management Science. 1990;**29**:437-449

[59] Alister C, Araya M, Becerra K, Saavedra J, Kogan M. Preharvest interval periods and their relation to fruit growth stages and pesticide formulations. Food Chemistry. 2017;**221**:548-554

[60] Balsari P, Marucco P. Sprayer adjustment and vine canopy parameters affecting spray drift: The Italian experience, In: Proceedings, International Conference on Pesticide Application for Drift Management, Washington, DC; 1989. p. 109-115

[61] Huo R, Salazar JD, Hyder K, Xu XM. Modelling non-systemic pesticide residues in fruits with initial deposit variability and weather effects. Food Additive and Contaminant. 2007;**24**:1257-1267

[62] Marin A, Oliva J, Garcia C, Navarro S, Barba A. Dissipation rates of cyprodinil and fludioxonil in lettuce and table grape in the field and under cold storage conditions. Journal of Agricultural and Food Chemistry. 2003;**51**:4708-4711

[63] Perrin RM, Wege PJ, Foster DG, Bartley MR, Browde J, Rehmke A, Scher H. Fast release capsules: A new formulation of lambda-cyhalothrin. Proceedings British Crop Protection Conference—Pests and Diseases. 1998;**1**:43-48

[64] Perić P, Dimić N, Stamenković S, Marčić D. Efficacy of lambda and gamma-cyhalothrin in *Cydia pomonella* L. and aphis pomi Deg. control. Pesticidi i fitomedicina. 2004;**19**:97-104

[65] Sims SR, Appel AG, Eva MJ. Comparative toxicity and repellency of microencapsulated and other liquid insecticide formulations to the German cockroach (Dictyoptera: Blattellidae). Journal of Economic Entomology. 2010;**103**:2118-2125

[66] Pogoda MK, Pree DJ, Marshall DB. Effects of encapsulation on the toxicity of insecticides to the oriental fruit moth (Lepidoptera: Tortricidae) and the predator Typhlodromus pyri (Acari: Phytoseiidae). The Canadian Entomologist. 2001;**133**:819-826

[67] Elbert A, Nauen A. New applications for neonicotinoid insecticides using imidacloprid as an example. In: Rami A, HorowitzIsaac I, editors. Insect Pest Management. Berlin, Heidelberg: Springer; 2004. pp. 29-44

[68] Eichner R. Imidachloprid in the Product Confidor Insecticide. ACT: National Registration Authority for Agricultural and Veterinary Chemicals; 1994. 23 p

[69] Elbert A, Nauen R, Leicht W. Imidacloprid, a novel chloronicotinyl insecticide: Biological activity and agricultural importance. In: Ishaaya I, Degheele D, editors. Insecticides with Novel Modes of Action. Springer Berlin Heidelberg. 1998. pp. 50-73

[70] Gupta S, Gajbhiye VT, Agnihotri NP. Leaching behavior of imidacloprid formulations in soil. Bulletin of Environmental Contamination and Toxicology. 2002;**68**:502-508

[71] Sarkar MA, Biswas PK, Roy S, Kole RK, Chowdhury A. Effect of pH and type of formulation on the persistence of imidacloprid in water. Bulletin of Environmental Contamination and Toxicology. 1999;**63**:604-609

[72] Sarkar MA, Roy S, Kole RK, Chowdhury A. Persistence and metabolism of imidacloprid in different soils of West Bengal. Pest Management Science. 2001;**57**:598-602

[73] Rasero FS. Características de los formulados comerciales de plaguicidas. Influencia sobre los residuos encontrados en alimentos. In: Valverde A, editor. Residuos de Plaguicidas ′96. IV Seminario Internacional sobre residuos de plaguicidas, Almería, España: PRRG, Universidad de Almería. 1998. pp. 125-140

[74] FAO. Manual sobre Elaboración y Empleo de las Especificaciones de la FAO para Productos Destinados a la Protección de las Plantas. 5ta ed. 1999. 167 p

Chapter 4

Evaluation of Insecticides in Protective Clothing

Melina Espanhol Soares and Flávio Soares Silva

Additional information is available at the end of the chapter

http://dx.doi.org/10.5772/intechopen.76075

Abstract

The exposure to insecticides causes several health problems, which can be aggravated by more toxicity. Therefore, to avoid this exposure, it is required to use protective clothing. The use of protective equipment against pesticides is indispensable and essential from the preparation/handling regulations of the pesticides spray to the application of diluted formulations. However, even with this protection, workers are not totally immune to the contamination of pesticides. There are several factors that contribute to the loss of efficiency of protective clothing against pesticides, such as field use, activity of application, the type of material, seam presence, clothing model, types of formulation used in the application, the process of washing, and the ironing of clothes after their use.

Keywords: insecticide, protective clothing, agrochemicals, exposure, contamination

1. Introduction

The extensive exposure of insecticides, mainly from organophosphates (OPs), organochlorines, carbamates, and pyrethroids, causes several damages to the worker health, such as poisoning, neurotoxicity, cancer, and leukemia [1]. The exposure to pesticides can occur through the dermal and respiratory tract, by direct contact such as the application of pesticides in the control of pests and weeds, the handling of the formulation, the transportation of the products, and even in the removal of the protective clothing after its use, and by indirect contact such as re-entry into the culture after the application of pesticides and washing contaminated clothing [2].

The greatest risk of poisoning by agrochemicals occurs due to the lack of use of protective equipment, and as a consequence, serious diseases such as psychiatric disorders and respiratory problems are caused [3]. In general, the use of protective equipment is commonly used by agricultural production organizations, where there is a greater oversight by the government, and the producers are better instructed in agrotoxic applicators area [4].

The use of protective equipment by rural workers is essential in Brazil, as can be seen in NR 6 [5]. However, it is common knowledge that the equipment marketed in Brazil, often supplied to workers by their employers according to the legislation, do not have adequate efficiency, exposing their users throughout the working day [6–9].

The manufacturer is responsible to sustain the quality of the personal protective equipment (PPE), which gave source to the Certificate of Approval (CA). Also, the producer should provide the information about the maximum permitted cleaning and hygiene procedures and indicate, when needed, the number of hygienization above which it is necessary to revise or replace the equipment, in order to guarantee the original level of protection [4].

To be marketed in the country, PPE must have the CA issued by the Ministry of Labor and Employment (MTE) and meet the pesticide protection requirements of International Organization for Standardization—(ISO) 27065: 2011 [10]. For the penetration test, the ISO 22608 (Protection Against Liquid Chemicals—Measurement of repellency, penetration and penetration of liquid pesticide formulations through protective clothing materials) is used [11]. In Brazil, the solution used in the tests is the herbicide Roundup Original®, classified as a soluble concentrate with 480 g/L glyphosate isopropylamine salt (48%) to replace the 5% solution of the pendimethalin active ingredient established in the procedure of ISO 22608: 2004 [5–7].

The 27,065 establishes that the evaluation of protective clothing must begin with materials and seam tests [10]. For impermeable materials, the ISO 6529 (determination of resistance of protective clothing materials to permeation by liquids and gases) is used [12]. In the permeation test, the chemical moves through the protective material by means of the molecular diffusion process with water, which is situated on the other side of the test cell [12].

The factors that may interfere in the permeability of pesticide penetration in the protective materials are air permeability ($cm^3/cm^2/s$), water vapor transmission rate ($g/h\ m^2$), viscosity (η), stress (mN/m) of pesticides, and characteristics of the protective material [13].

The characteristics of the protective materials can be damaged through washing procedure, which influences the protective efficiency of the garments. The washing procedure causes breakage in the fibers and tissues due to mechanical agitation of the washing machines, water, temperature, the number of wash cycles, and tissue drying [14]. Fabrics with a higher coat of fluorochemicals can withstand more than 20 washes, depending on the type of wash [15].

The evaluation methods of the whole body of garments to agrochemicals are grouped into three categories: qualitative, semiquantitative, and quantitative [7]. Qualitative methods are usually based on visual observations of the presence or absence of dermal exposure labeled with colored or fluorescent pigments in accordance with procedures established by the American Society of Testing and Materials (ASTM) described in ASTM F 1359: 2007 [16] and ISO standard 17491-4: 2008 [17]. Quantitative methods are performed by quantifying the pesticides themselves or markers added to the sprayed grouts. Quantitative or qualitative methods are based on the penetration of pesticides, dyes, or traces added to the syrup and porous materials or possible openings in clothing.

The efficiency of the protective clothing involves the choice of material, design, field performance testing, and efficacy tests with various types of pesticides. This way, the field worker can use the clothing sanitizing for other types of pesticides, and not just one [7].

This chapter presents a brief review of the dermal exposure of pesticide applicators/manipulators, as well as the efficiency of the protective clothing used in this activity and the factors that undermine this efficiency.

2. Dermal exposure assessments to insecticides

2.1. Field evaluation

Field evaluation involves the predominance of dermal exposure measurements in the application of agrochemicals, whose purpose is to verify the performance of the spray system in contact with the crop and the applicator [18].

The whole-body dosimetry method documented by the Organization for Economic Cooperation and Development (OECD) is widely used in the evaluation of protective clothing in the Exposure to Pesticides during Agricultural Application in compliance with the Guidance Document for the Conduct of Studies of Occupational [18]. This method is an alternative to the patch method and uses cloth layers to measure the body exposure underneath protective clothing. Parts of the head or hands may be worn for the evaluation of dermal exposure [19].

Exposure levels can be expressed as mL of spray deposited on each body part per hour of application (considering the spray concentration and the application time). The exposed parts depend on several factors, such as spray drop size, greenhouse or open-field application, spray type, spray culture structure, and the type of formulation used in application [19].

Frenich et al. evaluated the dermal exposition with the whole-body dosimeter method of spraying with fenitrothion, methidathion, malathion, dimethoate, chlorpyrifos-methyl, and methamidophos under greenhouse conditions. Then, they checked that the legs were more exposed, and which fine droplets of spray increase the dermal exposure by the spray body [20].

The patch method has also been used to assess the dermal exposure of pesticides in different parts of the worker's body [21]. Leme et al. [22] investigated the dermal exposure and malathion penetration inside the dressing using patch method prior to each nebulization and placed absorbents under and on EPI dress in the chest and upper chest wall (back) (80% of the samples contaminated with malathion).

To evaluate the exposure of tractor operators during the application of fenitrothion (organophosphate) in apple cultivation in southern Brazil, absorbents were used on the protective clothing of the artificial operator, quantifying the exposure on the clothing with values below 0.18 mg/kg [23].

The efficiency of two water-repellent personal protection was 96.7 and 96.2% for the tractor-sprayer in turbot sprays with turbopulverizer. The potential dermal exposures (in the most exposed areas of the spraying tractor) in the descending order were the feet, arms, thighs + front legs and trunk-back [24].

Goede et al. assigned scores to assess the effect of factors determining occupational exposure in order to correct and classify the dermal exposures of body parts. In tank preparation and tank-filling activities, dermal exposure in the hands is greater (per unit area) than in other parts of the worker's body [25]. The determinants of occupational exposure were viscosity and volatility of the applied substance, particle type, temperature during application, droplet size, particle size of the applied substance, the type of work performed (manual or automated), and was analyzed if the spray reaches the upper or the lower part of the body [26].

Some factors may increase the exposure of applicators in the field, as the structure and height of the crop increasing the spray volume applied in the same type of nozzle used in spraying, spray angle in relation to the worker [27], and correct worker clothing at the time of application [19].

Another exposure factor studied by Kasiotis et al. [28] is the re-entering of the treated crops, in which such exposure can vary according to the tasks of tying or pruning (pesticide residue can be transferred from the foliar surface of a plant to the worker). In this study, the difference in the exposure of SC insecticide (tebufenozide) and an EC fungicide (bupirimate) in tomato and pepper crops was verified. Workers' dermal exposure in applications of insecticide malathion at greenhouse pepper culture was higher in the upper body of the worker for water-repellent cotton, cotton/polyester, and cotton garments. This fact occurred due to spraying directed toward the top and toward the aerial part of the plants cultivated in bench [29].

Through the assessment of dermal exposure with different spraying equipment, it is possible to classify the risk conditions (safe or unsafe) for the pesticides used, based on the Noel ratio of the substance used in relation to the exposure dose of the applicator [30].

2.2. Laboratory evaluation

In Brazil, studies to evaluate the exposure and protection offered by PPE dressing have recently started and have been of great importance, since they analyze clothing marketed under local exposure conditions and can help in specific standards and tests according to the need of the country.

The evaluation of protective clothing against pesticides uses methods of dermal exposure assessment grouped into three categories: qualitative, semiquantitative, and quantitative. Qualitative methods are usually based on visual observations of the presence or absence of dermal exposure labeled with colored or fluorescent pigments in accordance with procedures established by the American Society of Testing and Materials (ASTM) described in ASTM F 1359:2007 [16] and ISO 17491-4:2008 [17].

Quantitative methods are performed by quantifying the pesticides themselves or markers added to the sprayed grouts. The quantitative evaluation described in ISO 16602 makes it possible to classify sets of protection against chemical substances and determine the useful life of these garments [31]. The process of evaluating the efficiency of the PPE starts with the selection of the materials that will be used in the manufacture of the set in laboratory tests.

International standards establish ways to evaluate the effectiveness of PPE dressings against agrochemicals. As of September 2009 in Brazil, the Ministry of Labor and Employment (MTE) by Ordinance No. 121/2009 established methods for the evaluation of these garments in relation to the repellency, permeation, and penetration of pesticides according to ISO 27065: 2011 [10], used internationally [32].

The type of the material and the seam of the parts determine the level of the PPE according to the protection requirements standard ISO 27065: 2011 [10]. The ISO 27065 sets that garments made are four levels: 1a, 1b, 2, and 3 against pesticides, and tests must be carried out on material, sewing, and whole garment [10].

In this requirement standard (ISO 27065), depending on the performance of materials and seams, full protection sets, or full-body garments and porous materials, the protector kits are classified in levels 1b or 2. According to this standard, for the sets to be classified at level 1b, the penetration of the test substance in the material and at the seam should be less than or equal to 40%, and for level 2, less than or equal to 5%, evaluated with the procedure of ISO 22608:2004 [10]. In this standard of ISO 27065:2011, requirement tests and criteria for minimum protection of materials, seams, and the complete sets themselves to assess the minimum safety and classify PPE against agrochemicals are defined.

The determination of the clothes classified at level 2 in performance is made with porous materials and seam with needles and thread, and level 3 is made with non-porous materials and welded seams, impermeable, as established in the norm of requirements [10].

Table 1 establishes the criteria for the tests carried out with protective clothing for material types (level) according to ISO 27065.

	Specific performance test	**Level**
Material requirements	Liquid penetration resistance (ISO 22608)	1b and 2
	Resistance to penetration by liquid under pressure (ISO 13994 Method E)	3
	Resistance to permeation (ISO 6529 Method A)	3
	Tensile strength (ISO 13934-1)	1, 2, and 3
	Tear strength (ISO 9073-4	1, 2, and 3
Seam requirements	Seam penetration resistance (ISO 22608)	1b and 2
	Resistance to penetration by liquid under pressure (ISO 13994 Method E	3
	Resistance to permeation (ISO 6529 Method A)	3
	Tensile strength (ISO 13934-1)	1, 2, and 3
	Tear strength (ISO 9073-4	1, 2, and 3
Whole-garment requirements	Practical performance test	1, 2, and 3
	Low-level spray test (ISO 17491-4 Method A)	2
	High-level spray test (ISO 17491-4 Method B)	3

Table 1. Requirement tests for level 1, 2, and 3 garments.

The protection criteria established in the ISO 27065: 2011 for PPE classified in level 2 [10] are materials and seams—penetration of <5% of the aqueous solution containing 5% of pendimethalin or glyphosate, evaluated with ISO 22608: 2004 standard, to the assessment of whole-body PPE (using automated closed chamber)—1-cm diameter stains of methylene blue dye, evaluated using the procedure of ISO 17491-4: 2008 [17]. The criteria for PPE at level 3 (impermeable material) are normalized in the final breakthrough time of ≥30 min: the final breakthrough time is standardized when the permeation rate normalizes at 1.0 $\mu g/cm^2$ min [10].

2.2.1. *Toxicology and risk assessment*

The registration of insecticides or other agrochemicals is a complex process and goes through several stages, including biological tests with animals and microorganisms. This way, it is possible to know the causes of acute or chronic exposure to those who manipulate organisms affected by exposure. In **Figure 1**, a layout is described, which shows the registration of a new pesticide.

The efficiency tests to protective clothing against pesticides performed by ISO 22608 and according to ISO 27065 must have penetration to pesticides equal to or less than 5% and a herbicide with a low toxicological class was used (glyphosate) [10]. However, in these studies,

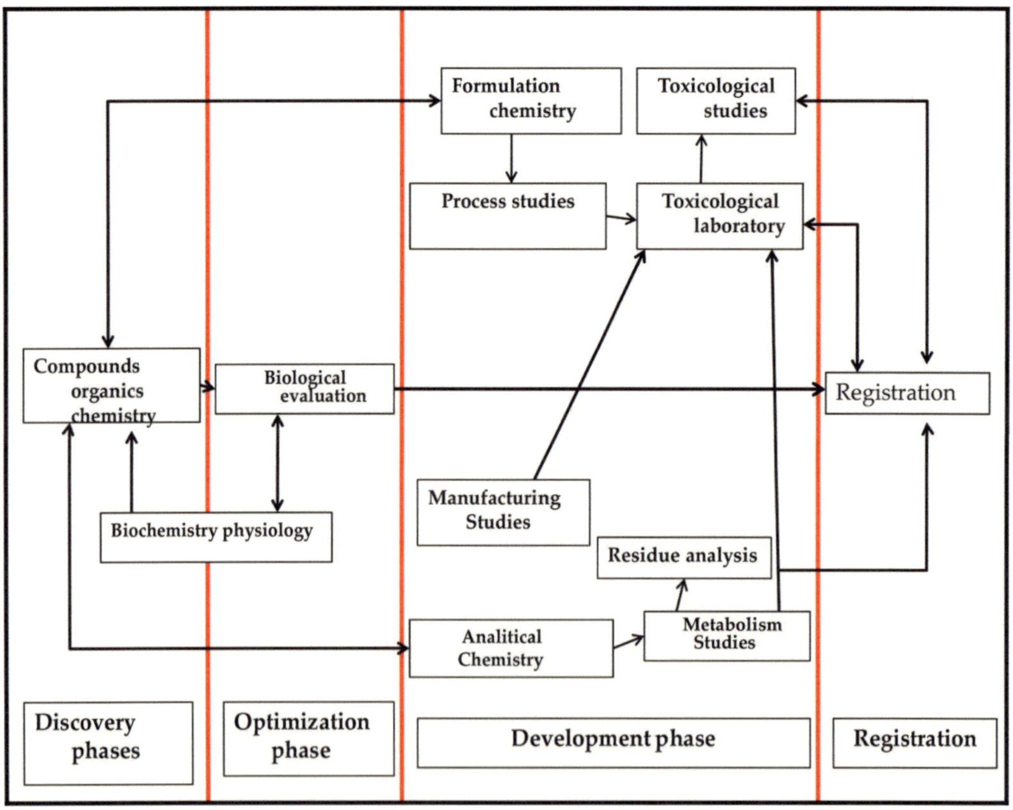

Figure 1. New pesticide registration process [33].

only the penetration of pesticide solutions is considered and not the pesticide toxicology. In Brazil, several agrochemicals are still commercialized, which have medium and high toxicity, such as the organophosphorus, organotin, chloroaromatic, and others. The margin of exposure (MOE) analysis is a tool to assess the risks in the exposure to pesticides, and animal toxicity tests are compared with levels of human exposure. The NOEL value is 500 mg/kg/day for glyphosate according to USEPA [34]. To obtain a safety margin of exposure, the MOE is linked to NOEL in relation to the exposure dose ratio of pesticides. Some examples of MOE are chlorpyrifos, which have a value of 100 according to USEPA [35] and can change in 100–820 rate if the applicator is wearing protective clothing. To the pesticide profenos, the MOE value is 300 [36]; to the acephate, the MOE (calculated as acute human NOEL of 1 mg/kg/day/estimated human exposure) is <10, depending on the type of spraying used [37].

Thus, the calculation of the safety margin of exposure for the penetration of 5% active ingredients and 16 cm^2 area according to ISO 22608, the MOE value would be 0.5 mg. For acephate, which is an active ingredient more toxic than glyphosate, and assuming that the worker's body area is 21,050 cm^2 as defined by Nuyttens et al. [38], the ingredients actively penetrated would be defined by 657,81 mg (considering 1 day of work). This value is divided by the middle-weight (70 kg) results in condition mg/kg/day. The NOEL value divided by the exposure value calculated previously results in MOE, which is 0.7. This value is low, but it is considered as risk for workers.

Compared with the glyphosate risk used in the tests in Brazil, and using the same body area and the mean body weight data, but with NOEL of 500 mg/kg/day, and the mol of 52.46, a value below that recommended in the work by Lake [39] featured MOE for glyphosate applicators in a value of 83.

Researches about the potential risk of exposure of pesticide applicators have been studied in some European countries, in which the exposure was based on the Agricultural Operator Exposure Model (AOEM) by collecting the data in mg/person of the ingredients and comparing with Acceptable Operator Exposure Levels (AOELs) [40].

Most of the water-repellent materials found in the current market are treated with fluoride and carbon polymerization substances (**Figure 2**), combined in fluoropolymers, giving a high water repellency characteristic to the material [41].

The efficiency evaluation of protective clothing was performed in a laboratory by a closed chamber, which provides less variability in relation to the evaluation in field as environmental conditions and structure of crop. To evaluate the protective clothing in a laboratory, Espanhol-Soares et al. used tracers under field conditions in sugar cane culture and a dummy dressed in sampler clothes in an automated closed chamber. Also, a protective equipment was used to get a greater penetration for evaluation in camera. The penetration of the tracer in the clothes evaluated in the chamber was higher than in the field, ranging from 3.2 to 24% for 0–30 uses and washes, respectively [42].

Machado evaluated the efficacy of whole-body PPE used in the applications of insecticide malathion, for the control of the dengue mosquito. It has been checked that garments washed

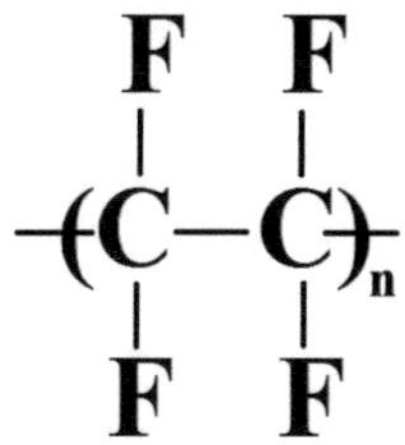

Figure 2. Chemical structure of polytetrafluoroethylene polymer. C, carbon; F, fluor; and n, repeating units (polymer).

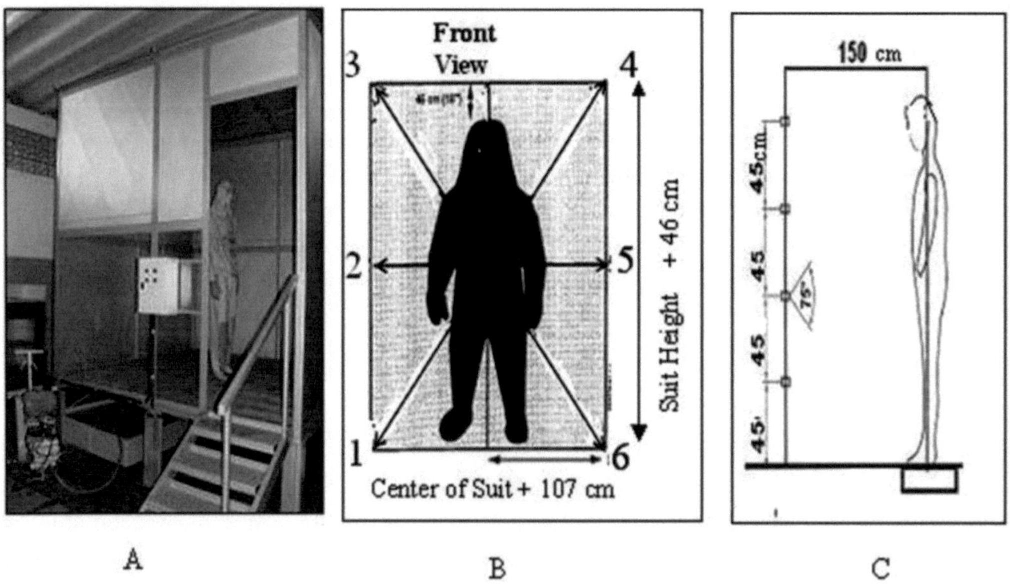

Figure 3. (A) A closed chamber used for assessment of protective clothing against pesticide. (B) Spraying layout according to ASTM 1359. (C) Spray layout (ISO 17491-4) [43].

under field conditions show a lower efficiency of clothes washed in the laboratory past and not only passed with 10 and 20 wash cycles, although the values are higher than 98%. Under laboratory conditions, the 30 cycles of washes and hot ironing do not significantly reduce the efficiency of PPE [6]. This result is due to the washing procedure number 8 - called gentle, established in ISO 6330: 2000 [6].

Espanhol-Soares et al. investigated the dermal exposure using a protective clothing applying the efficiency tests conforming to ASTM F 1359, under a spray bath using an exposure level greater than ISO 17491-4. The ISO 17491-4 procedure is required to evaluate complete protection sets in the standard ISO 27065. However, in this work, the ASTM standard was adapted to the flow similar to that of ISO 17491-4 for method A. Nevertheless, the total dermal exposure without the use of the protective clothing was 21426.5 mL L^{-1}, according to ASTM 1359 (total volume sprayed 1.98 L). This value is higher than 2265.3 mL L^{-1}, obtained for method B

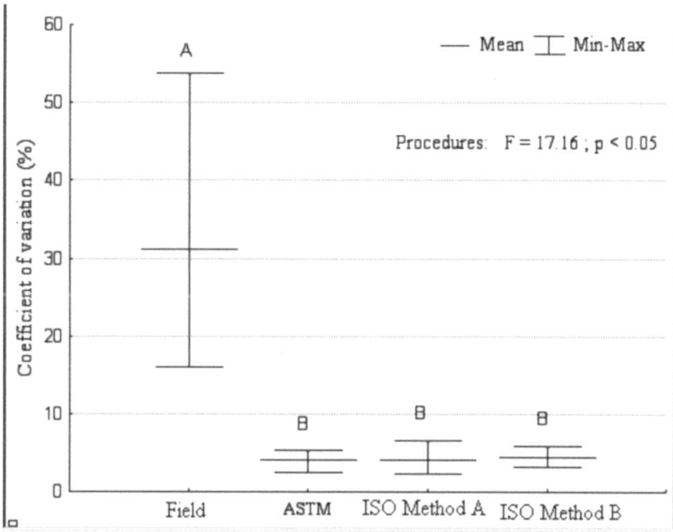

Figure 4. Percentage and variation coefficient of potential exposure in the field and the dummy according to modified ASTM F1359, methods A and B by ISO 1749-4 [43].

(total volume sprayed 4.56 L) and 587.1 mL L^{-1}, to the total exposition according to ISO 17491-4, using method A (total volume sprayed 1.88 L). **Figure 3** shows the spray methodology differences between ASTM 1359 and ISO 17491-4 [43]. These results imply in the comparison with the best methods to be chosen in the evaluation of clothing. In this work, a tracer was used, although the methods took into account the interfacial tension of the sprayed solution, but there may be differences if these tests were carried out with insecticides or other agrochemicals, due to their characteristics of the formulated product.

Espanhol-Soares et al. [42] evaluated the differences of dermal exposures in new protective clothing in field and laboratory conditions. The use of tracers in the evaluations enabled to obtain a coefficient of variation (CV%) in the field of 27%, and the variation for the dermal exposure in the laboratory using the dummy was between 3 and 4%, as shown in **Figure 4** [43].

3. Loss of efficiency

3.1. Pesticides (types of formulations, active ingredient, viscosity, etc.)

The penetration of the pesticides is defined by the passage of a chemical compound, which passes through openings, pores of the materials, seams, holes, or other imperfections of overlapping of the clothing [43]. The movement of the pesticide molecule occurs in three stages: adsorption, absorption, and desorption. The adsorption is controlled by the ratio of the tissue surface energy to the surface tension of the spray mixture. Absorption is the way the pesticide interacts within the tissue structure. This movement is determined by the tissue structure and the size of the pesticide molecule. Desorption is the ability of the pesticide to penetrate the tissue [44].

The pesticides used in the ISO 22608 test, which were diluted and undiluted, are glyphosate, chlorpyrifos, and copper hydroxide in a water-repellent material, being greater penetrations for diluted solutions, mainly for solutions of chlorpyrifos and copper hydroxide. For the insecticide chlorpyrifos, the material was rejected after five uses and washes with the value of penetration of 5.5% for the diluted solution. For the solution undiluted material, it was disregarded after 20 washes of the material used in the field, but the evaluation with the glyphosate solution used in certification tests of garments was only reproved after 30 uses and washes after use in the sugar cane culture. These results indicate that it is necessary to carry out the tests with insecticides also for the clothing certification [45].

The penetration and retention of pesticides depend on the type of material: cotton (100%) retains more atrazine than other materials, due to the strong attraction of the molecules in tissue [46]. Nelson et al. evaluated that the differences on the retention of pesticides in protective materials are attributed to the type of the fiber of the material, since the retention of pesticides in cotton materials (100%) occurs in a greater proportion than in cotton/polyester materials (50/50%). The retention of carbaryl, in the formulation-concentrated suspension, and atrazine (concentrated suspension and wettable powder) is higher in cotton material; the cypermethrin (wettable powder) and the trifluralin (emulsifiable concentrate) are higher in cotton/polyester material [47].

3.1.1. Insecticides impregnated in protective clothing

In addition to the evaluation of insecticides in protective clothing, the protective clothing has been impregnated by insecticides using the coating method [48]. Protective clothing used in areas with a higher incidence of diseases such as malaria, Chikungunya, dengue, Yellow fever, African tick-bite fever, *Aedes aegypti,* and Culex mosquitoes are generally impregnated with permethrin. The mixture of this insecticide with repellents may be 100%, depending on the applied dose and the type of coating applied on the clothes (as dip coating or spray). However, the use and wash decrease the efficiency, and it is suggested that after five uses, the insecticides must be reapplied in clothing [49].

The insecticide permethrin was impregnated in military uniform clothing (65% cotton and 35% polyester, weight of 220 g/m^2) for prevention in malarious areas. After its use in the field, the residual concentration of permethrin is ≥200 mg permethrin/m^2. The insecticide-coated clothing after 218 washes obtained the remaining permethrin at a concentration of 130 and 95 mg/m^2. The established value for *A. aegypti* mosquito mortality was 200 mg/m^2 [50]. Clothing impregnated with a mix of repellent and organosorption inhibited 56.25% of bites. A group of the clothes were manually impregnated with the repellent KBR3023 (10 g/m^2) and another group were impregnated through the combination of pyrimiphos-methyl (150 mg/m^2).

3.2. Washing and permeation processes

An important factor in the loss of repellency of the protective materials against pesticides is the water temperature and the movements during washing used in garments. These facts affect the efficiency of cleaning agents and affect the protection of cotton fabrics as well [51].

The water repellency and the contact angle with the Teflon-treated polyester increase with the concentration of water-repellent substances applied to the fabric and decrease with the increase in the number of AATCC washes of 22 [52].

Obendorf et al. evaluated the adsorption of ionic surfactants present in soaps after washing processes in cotton fabrics. They found that the change in pH in the wash solution affects the adsorption of surfactants in the tissues. Cationic surfactants are adsorbed on cotton fabrics because of the negative charges [53]. This mechanism of adsorption may explain why certain pesticides are more retained in the tissues.

3.3. Design of the garments (type of material, presence of sewing, weight, etc.)

For many times, the inefficiency of protective clothing is due to improper use or poorly constructed, and pesticides penetrate into clothing through openings, seams, zippers, folded sleeves, and poor overlaps of sleeves with gloves [54]. Such inefficiency of protection against pesticides may be absorbed by the skin of the worker [54].

The resistance to penetration of protective material against pesticides depends on the methodology used and the type of tissue. In fabrics made of 100% cotton with a water-repellent treatment (Phobol oil), penetrations below 1.6% have been found when evaluated by the trough, pipette, and atomizer test method. However, in cotton/polyester composite fabrics, the penetration is 12.8% with the pipette method, 16.5% with the gutter method, and no penetration with the atomizer method [42].

Although water-repellent fabrics cause discomfort to workers, especially in the hottest agricultural regions, it is believed that with the proof of EPI efficiency, the applicator distribution on the regions of the body, and with the methodologies of evaluation of the exposures, the EPI should be recommended for the most exposed areas of the body and provide safety and lesser discomfort to the worker as well [55].

Protective clothing has been studied through the ergonomic property testing in dummies in chambers with controlled environmental and exposure conditions [56]. In the dermal exposure, dummies can be used for assessment methods. Therefore, the penetration or retention of insecticides is evaluated, as in the evaluation of malathion spraying on protective clothing used to control dengue [22]. However, there are few studies in the literature that quantitatively evaluate the efficiency of closed-loop protection sets.

Machera et al. used the procedure of ISO 22608 to evaluate the penetration of pesticides in protective clothing materials. For materials containing cotton/polyester (50/50%) with 215 g/m^2 treated with NanoTex® water repellent, the penetration was 2.4% after 15 washes. However, in cotton-dressing materials without water-repellent treatment with 287 g/m^2, the penetration was 18.7%, after five washes [29]. Therefore, it has been found that the cotton yarns in the material provide the highest penetration of the test solution.

Shaw and Schiffelbein tested approximately 100 different fabrics used in the manufacture of pesticide applicator clothing and verified that the highest levels of protection were found on water-repellent garments [9].

Oliveira and Machado Neto evaluated the penetration of the insecticide methamidophos into two types of tissues: a cotton water repellent (153 g/m^2 to 0.25 mm) fabric and another jeans fabric (458.66 g/m^2 to 0.75 mm). The authors noted that after 30 washes at manual washing with soap, more insecticides penetrated into lighter tissue (21.05%) than in the heavier tissue (0.12%) [57].

Marinho [7] evaluated the material of the protective clothing by ISO 22608 with seven types of washes with a machine programmed according to ISO 6330 and a manual washing as well. As a result, before the washing, all materials met the criterion of approval of the requirement rule ISO 27065 (ISO, 2011). After the 5, 10, 20, and 30 wash cycles, the penetrations of the test formulation (glyphosate) were higher than 5% for the materials without ironing process and no longer meeting the criteria of approval. All materials (cotton 100%, cotton 65% + polyester 35%), when ironing, had a penetration of <5%, in five uses and washes. However, the 50% cotton +50% polyester material had the same result in 20 uses and washes.

4. Conclusion

It is noted, according to the study, that there are still several gaps to be studied regarding exposure to insecticides and other pesticides. Protective clothing tested according to international standards uses only one type of pesticide (pendimethalin or glyphosate), which do not emphasize the toxicity of the product in contact with the skin of the worker. The means of exposure to insecticides can occur without the use of clothing, but the exposure also occurs with the use of protective clothing. The loss of efficiency of clothing can be due to wear by insecticide applicators, washing, the use of soap at the time of washing, the presence of seams, and improperly made openings. Another exposure factor is the use of clothing by different types of pesticide formulations, which ensures wear. In addition, the factor discussed is the quality of the clothes that are put up for sale; even the certified clothing used in the application of insecticides does not present adequate exposure to the worker throughout the workday. There are differences in results between the tests carried out with the clothes in the laboratory and in the field, mainly due to wear factor by the use and contamination with other types of formulations that interact with the fabric or material of the dress. It is important to emphasize that many insecticides with high toxicity are still commercialized in the world, although many countries already prohibit their commercialization, as some organophosphates and organochlorines. This way, it is important to evaluate protective clothing with these types of pesticides.

Author details

Melina Espanhol Soares* and Flávio Soares Silva

*Address all correspondence to: melespanhol@yahoo.com.br

Universidade Federal de Itajubá (UNIFEI), Itajubá, MG, Brazil

References

[1] Keifer MC, Firestone J. Neurotoxicity of pesticides. Journal of Agromedicine [Internet]. 2007;**12**(1):17-25. DOI: 10.1300/J096v12n01_03

[2] Nicolopoulou-Stamati P, Maipas S, Kotampasi C, Stamatis P, Hens L. Chemical pesticides and human health: The urgent need for a new concept in agriculture. Frontiers in Public Health. 2016;**4**:148-153. Available from: http://journal.frontiersin.org/Article/10.3389/fpubh.2016.00148/abstract

[3] Faria NMX, Fassa AG, Facchini LA. Pesticides poisoning in Brazil: the official notification system and challenges to conducting epidemiological studies. Ciência & Saúde Coletiva. 2007;**12**(1):25-38. Available from: http://www.scielo.br/scielo.php?script=sci_arttext&pid=S1413-81232007000100008&lng=pt&tlng=pt

[4] Rosa BT, Borges LAC, Pereira SP, Antonialli LM, Chalfoun SM, Baliza DP. Estudo sobre boas praticas agricolas em uma associaçaoo de cafeicultores familiares por meio da análise de clusters. Coffee Science. 2017;**12**(1):49-59

[5] BRAZIL. Ministério do Trabalho e Emprego. Portaria no 3.214, de 8 de junho de 1978, Ministério do Trabalho e Emprego. Aprova as Normas Regulamentadoras – NR – do Capitulo V, Titulo II, da Consolidação das Leis do Trabalho, relativas a Segurança e Medicina do Trabalho. Norma Regulamentadora Equipamento de Proteção Individual-NR 6. Disponível em: http://trabalho.gov.br/images/Documentos/SST/NR/NR6.pdf [Accessed: 2018-01-10]

[6] Machado AA. Efficiency of individual protection equipment with a certificate of approval. against aquosa and oleosa calle malathion used in nebulization [thesis]. Jaboticabal: São Paulo State University; 2017

[7] Marinho MO. Loss of effectiveness of hydro-repellent materials in the protection against pesticides by washing process [dissertacion]. Jaboticabal: São Paulo State University; 2013

[8] Espanhol-Soares M. Determination of efficiency and useful life of individual protection sets to pesticides according to the procedure for use and wash [thesis]. Jaboticabal: São Paulo State University; 2012

[9] Shaw A, Schiffelbein P. Protective clothing for pesticide operators: Part II – Data analysis of fabric characteristics. International Journal of Occupational Safety and Ergonomics. 2016;**22**(1):7-11

[10] ISO. International Organization for Standardization. ISO 27065: Protective clothing—Performance requirements for protective clothing worn by operators applying pesticides. Geneva; 2011. 23 p

[11] ISO. International Organization for Standardization. ISO 22608: Protective clothing protection against liquid chemicals: Measurement of repellency, retention, and penetration of liquid pesticide formulations through protective clothing materials. Geneva; 2004. 11 p

[12] ISO International Organization for Standardization. ISO 6529: Protective clothing—Protection against chemicals—Determination of resistance of protective clothing materials to permeation by liquids and gases. Geneva; 2001. 21 p

[13] Jain R, Raheel M. Barrier efficacy of woven and nonwoven fabrics used for protective clothing: Predictive models. Bulletin of Environmental Contamination and Toxicology. 2003;**71**(3):437-446

[14] Daroux FY, Carr DJ, Kieser J, Niven BE, Taylor MC. Effect of laundering on blunt force impact damage in fabrics. Forensic Science International. 2010;**197**(1-3):21-29

[15] Marek J, Martinková L. Protective clothing. In: Williams JT, editor. Waterproof and Water Repellent Textiles and Clothing. Elsevier Ltd; 2018. pp. 392-445

[16] ASTM. American Society for Testing and Materials F 1359. Standard Test Method for Liquid Penetration Resistance of Protective Clothing or Protective Ensembles under a Shower Spray while on a Mannequin. West Conshohocken: ASTM

[17] ISO. International Organization for Standardization. ISO 17491-4: Protective clothing—Test methods for clothing providing protection against chemicals—Part 4: Determination of resistance to penetration by a spray of liquid (spray test). Geneva; 2008. 9 p

[18] So J, Ahn J, Lee TH, Park KH, Paik MK, Jeong M, et al. Comparison of international guidelines of dermal absorption tests used in pesticides exposure assessment for operators. Toxicology Research. 2014;**30**(4):251-260

[19] Cao L, Chen B, Zheng L, Wang D, Liu F, Huang Q. Assessment of potential dermal and inhalation exposure of workers to the insecticide imidacloprid using whole-body dosimetry in China. Journal of Environmental Sciences (China). 2015;**27**(C):139-146

[20] Frenich AG, Aguilera PA, Gonzalez FE, et al. Dermal exposure to pesticides in greenhouses workers: Discrimination and selection of variables for the design of monitoring programs. Environmental Monitoring and Assessment. 2002;**80**:51-63

[21] Selmi G da FR, Trapé AZ. Health protection for rural workers: The need to standardize techniques for quantifying dermal exposure to pesticides. Available from: http://www.scielo.br/scielo.php?script=sci_arttext&pid=S0102-311X2014000500952&lng=pt&tlng=pt

[22] Leme TS, Papini S, Vieira E, Luchini LC. Evaluation of personal protective equipment used by malathion sprayers in dengue control in São Paulo, Brazil. Cadernos de Saúde Pública [Internet]. 2014;**30**(3):567-576. Available from: http://www.scielo.br/scielo.php?script=sci_arttext&pid=S0102-311X2014000300567&lng=pt&tlng=pt

[23] Barcellos M, Faletti MM, Madureira LA dos S, Bauer FC. Analytical evaluation of the protection offered by sealed tractor cabins during crop pulverization with fenitrothion. Environmental Monitoring and Assessment. 2016;**188**(12):660

[24] Tácio MB, Oliveira ML, Machado-Neto JG. Efficiency of new clothes waterrrepelent int the protection of the tractor-driver in pesticides spraying in guava orchads with the air assisted sprayer. Revista Brasileira de Fruticultura. 2008;**30**:106-111

[25] Goede HA, Tijssen SCHA, Schipper HJ, Warren N, Oppl R, Kalberlah F, et al. Classification of dermal exposure modifiers and assignment of values for a risk assessment toolkit. The Annals of Occupational Hygiene. 2003;**47**(8):609-618

[26] Warren N, Goede HA, Tijssen SCHA, Oppl R, Schipper HJ, Van Hemmen JJ. Deriving default dermal exposure values for use in a risk assessment toolkit for small and medium-sized enterprises. The Annals of Occupational Hygiene. 2003;**47**(8):619-627

[27] Foqué D, Nuyttens D. Effects of nozzle type and spray angle on spray deposition in ivy pot plants. Pest Management Science. 2011;**67**(2):199-208

[28] Kasiotis KM, Tsakirakis AN, Richard Glass C, Charistou AN, Anastassiadou P, Gerritsen-Ebben R, et al. Assessment of field re-entry exposure to pesticides: A dislodgeable foliar residue study. Science of the Total Environment. 2017;**596-597**:178-186

[29] Machera K, Tsakirakis A, Charistou A, Anastasiadou P, Glass CR. Dermal exposure of pesticide applicators as a measure of coverall performance under field conditions. Annals of Occupational Hygiene. 2009;**53**(6):573-584

[30] Machado-Neto JG. Safety measures for handlers/workers against herbicide intoxication risk. In: Price A, Kelton J, Sarunaite L, editors. Herbicides, Physiology of Action, and Safety [Internet]. Rijeka: InTech; 2015. Available from: http://dx.doi.org/10.5772/61464

[31] ISO. International Organization for Standardization. ISO 16602: Protective clothing for protection against chemicals—Classification, labelling and performance requirements. Geneva; 2007. 40 p

[32] BRAZIL, Ministério do Trabalho e Emprego MTE, Portaria 121 de setembro 2009. Avaliable from: <http://www.quepia.org.br/site/portaria/2010_1808/Portaria%20121.pdf>

[33] Tordoir WF. The development of safe chemical pestice. In: Kee C, Jeyaratnam S, editors. Occupational Health In National Development. 1990

[34] United States Environmental Protection Agency (USEPA). Reregistration Eligibility Decision (Red) Glyphosate. Office of prevention, pesticides and toxic substances. 1993. https://www3.epa.gov/pesticides/chem_search/reg_actions/reregistration/red_PC-417300_1-Sep-93.pdf [Accessed: November 9, 2017]

[35] US Environmental Protection Agency (US EPA) Human Health Risk Assessment Chlorpyrifos. Office of Pesticide Program, Health Effect Division. 2000. http://www.epa.gov/scipoly/sap/meetings/2008/september/hed_ra.pdf

[36] US Environmental Protection Agency (US EPA) Human Health Risk Assessment Profenofos. Office of Pesticide Program, Health Effect Division (7509C), 2000. https://www3.epa.gov/pesticides/chem_search/cleared_reviews/csr_PC-111401_16-Jun-99_051.pdf

[37] Gammon D. Califronia department of pesticide regulation. Acephate risk characterization document. 2008. Available from: http://www.cdpr.ca.gov/docs/risk/rcd/acephate.pdf

[38] Nuyttens D, Braekman P, Windey S, Sonck B. Potential dermal pesticide exposure affected by greenhouse spray application technique. Pest Management Science. 2009;**65**(7):781-790

[39] Lake R. Health risk assessment: Glyphosate: Prepared as part of a Ministry of Health contract for scientific services. 2014. https://www.esr.cri.nz/assets/HEALTH-CONTENT/MoH-reports/FW14022-Glyphosate-FINAL-28-Oct-2014.pdf

[40] Wong HL, Garthwaite DG, Ramwell CT, Brown CD. Assessment of exposure of professional agricultural operators to pesticides. Science of the Total Environment. 2018;**619-620**:874-882

[41] Barreto C, Chen JP, Desai I, Finegold S, George A, Hu M, Nan K, Otake R, Rao A, Smolen C, Yin L, Zhao D. Improving the hydrophobicity of fabrics with the use of phosphonic acids. 2013. Available from: http://www.drew.edu/wp-content/uploads/sites/99/Team2.pdf

[42] Espanhol-Soares M, Nociti LAS, Gonçalves Machado-Neto J. Procedures to evaluate the efficiency of protective clothing worn by operators applying pesticide. Annals of Occupational Hygiene. 2013;**57**(8):1041-1053

[43] Shaw A, Cohen E, Hinz T, Herzig B. Laboratory test methods to measure repellency, retention, and penetration of liquid pesticides through protective clothing part I: Comparison of three test methods. Textile Research Journal. 2001;**71**(10):879-884

[44] Leonas KK. The mechanism of pesticide transmission through apparel fabrics: A comparison of drop and spray exposure methodologies. Archives of Environmental Contamination and Toxicology. 1991;**20**(3):427-431

[45] Espanhol-Soares M, Teodoro de Oliveira M, Machado-Neto JG. Loss of effectiveness of protective clothing after its use in pesticide sprays and its multiple washes. Journal of Occupational and Environmental Hygiene. 2017;**14**(2):113-123

[46] Taylor SM, Kim CJ, Lombadi J, Lea SM. Pesticide residue distribution on protective clothing fabrics as determined by SEM micrographs and their image analyses, performance of Protective Clothing: Issues and priorities for the 2t. In: Nelson CN, Henry NW. Century: Seventh Volume, ASTM STP 1386. West Conshohocken, PA: American Society for Testing and Materials; 2000

[47] Nelson C, Laughlin J, Kim C, Rigakis K, Raheel M, Scholten L. Laundering as decontamination of apparel fabrics: Residues of pesticides from six chemical classes. Archives of Environmental Contamination and Toxicology. 1992;**23**(1):85-90

[48] Faulde MK, Nehring O. Synergistic insecticidal and repellent effects of combined pyrethroid and repellent-impregnated bed nets using a novel long-lasting polymer-coating multi-layer technique. Parasitology Research. 2012;**111**(2):755-765

[49] Alpern JD, Dunlop SJ, Dolan BJ, Stauffer WM, Boulware DR. Personal protection measures against mosquitoes, ticks, and other arthropods. Medical Clinics of North America. 2016;**100**(2):303-316

[50] Most B, Pommier de Santi V, Pagès F, Mura M, Uedelhoven WM, Faulde MK. Erratum to: Long-lasting permethrin-impregnated clothing: Protective efficacy against malaria in hyperendemic foci, and laundering, wearing, and weathering effects on residual bioactivity after worst-case use in the rain forests of French Guiana. Parasitology Research. 2017;**116**(2):677-684. DOI: 10.1007/s00436-016-5333-6

[51] Liu H, Wang Y, Gong RH, Zeng J, Ding X. The relationships between washing parameters, fabric movement, and wrinkling in a top-loading washer. Textile Research Journal [Internet]:40517517700197. Available from: https://doi.org/10.1177/0040517517700197

[52] Mortazavi V, Khonsari MM. On the degradation of superhydrophobic surfaces: A review. Wear. 2017;**372-373**:145-157

[53] Obendorf SK, Dixit V, Woo DJ. Microscopy study of distribution of laundry fabric softner on cotton fabric. Journal Surfactant Detergent. 2009;**12**:225-230

[54] Aprea C, Terenzoni B, De Angelis V, Sciarra G, Lunghini L, Borzacchi G, et al. Evaluation of skin and respiratory doses and urinary excretion of alkylphosphates in workers exposed to dimethoate during treatment of olive trees. Archives of Environmental Contamination and Toxicology. 2004;**48**(1):127-134

[55] de Oliveira ML, Machado Neto JG. Segurança na aplicação de agrotóxicos em cultura de batata em regiões montahosas TT—Safety of pesticides application on potato crop on sloped areas. Revista Brasileira de Saúde Ocupacional. 2005;**30**:15-25. DOI: 10.1590/S0303-76572005000200003

[56] Holmér I. The role of performance tests, manikins and test houses in defining clothing characteristics relevant to risk assessment. Annals of occupational hygiene [Internet]. 1999;**43**(5):353-356. Available from: http://www.ncbi.nlm.nih.gov/pubmed/10481635

[57] Oliveira ML, Machado-Neto JG. Permeability of two types of cotton fabric used in personal protective clothing to the insecticide methamidophos. Bulletin of Environmental Contamination and Toxicology. 2005;**75**:1156-1162

Chapter 5

Particulate Nanoinsecticides: A New Concept in Insect Pest Management

Teodoro Stadler, Micaela Buteler,
Susana R. Valdez and Javier G. Gitto

Additional information is available at the end of the chapter

http://dx.doi.org/10.5772/intechopen.72448

Abstract

Nanostructured alumina (NSA) has insecticidal properties and has been demonstrated to be effective against stored product insect pests in laboratory bioassays. NSA is a nano-engineered material synthesized by oxidation of metals, and resulting particles show fixed electric charges. On the other hand, insects exhibit their own electric charges generated by triboelectrification. We propose that the mechanism of action of NSA involves two steps occurring in sequential order. First, a strong electrical binding between negatively charged NSA particles and positively charged insect. Next, dehydration of the insect occurs due to the strong sorbtive action of the NSA particles that remove the insect cuticular, leading to death by dehydration. As postulated for insecticidal inert powder in generals, particles attach to the insect cuticle surface disrupting water balance, and effectiveness decreases as ambient humidity increases, given that electrostatic bond forces are reduced by electrostatic discharge. The high insecticidal efficacy of NSA is a result of its intrinsic electric charge, small particle size and high sorptive potential due to its large specific surface area. NSA could provide an alternative to conventional synthetic organic insecticides due to its strong insecticidal properties with the advantage that its mechanism of action involves physical and electrostatic phenomena.

Keywords: nanoinsecticides, mode of action, triboelectrification, *Sitophilus oryzae*, insecticide powders

1. Introduction

The advent of synthetic organic pesticides by mid-1950s made the control of insect pests highly effective and despite their drawbacks, most of these active principles are still used in modern agriculture. The use of synthetic insecticides has allowed an increase in yields and lowered the cost of farming. Synthetic organic pesticides also remain an important tool to control vectors of infectious diseases of both humans and domestic animals, leading to a great reduction in their incidence in many areas of the world. However, synthetic organic insecticides may impact negatively on human health and ecosystems, affecting populations of non-target organisms and biodiversity [1]. Moreover, the accumulation of active ingredients or their metabolites in the environment as well as in organisms, may lead to bioaccumulation, where these pollutants enter the food chain, posing a serious threat to both wildlife and humans [2, 3].

Caught in a vicious circle? Agriculture has waged a costly struggle fighting insects by constantly rotating obsolescent pesticides in a desperate strategy of chemical warfare. However, a comprehensive and successful strategy for minimizing acute and chronic risk from pesticide use should be based on research initiatives aimed at radical changes in pest management strategies and the replacement of the synthetic organic pesticides with effective but less hazardous substances [4]. Part of the research on new biorational pesticides focuses on natural products such as plant extracts, oils and inorganic products. These are frequently a source of new chemical classes of insecticides, as well as environmentally and toxicologically less hazardous active ingredients than many of the conventional products used for insect pest control. Furthermore, new active ingredients often have mechanisms of action or molecular target sites which still remain unexploited by conventional marketed pesticides [5]. Hence, substances with new properties are promising tools for crop protection and food production, opening new frontiers in pest management [6]. However, only 14% of the pesticides on the market are biorational products and only 1% consists of natural products like plant extracts, essential oils and insecticide powders [7].

1.1. Nanotechnology as a source of modern pesticides

Nanotechnology is a collective term for a wide range of technologies that deal with structures and processes at the nanometer scale. The transition to the nanometer scale (10^{-9} m), leads to an increase in dominance of quantum-physical effects, optical, magnetic, electronic, mechanical and chemical properties [8]. Because of its potential for the fundamental transformation of entire technology fields, nanotechnology will not only influence technological development in the near future, but also have economic, ecological and social implications [9]. The size reduction to the nanometer range often leads to characteristic properties of materials which are useful for new applications and which do not occur in the case of macroscopic pieces of the same material. These include, for example, higher breaking strength at low temperatures as well as superplasticity at high temperatures, formation of additional electronic states, high chemical selectivity of the surface structures and a markedly increased surface energy [10].

Nanotechnology has advanced rapidly over the last 10 years and numerous nanomaterials, with a variety of potential applications, have been developed. For instance, improvements in medical science through nanotechnology offer the possibility to develop novel diagnostics and therapeutics [11], as well as new nano-engineered products with pesticide properties which have shown to be promising as tools for low impact or alternative organic agriculture and food production [12–15]. Engineered versions of conventional pesticides, growth regulators and seed treatment agents are among the first nano-chemicals that could be used in agriculture [13]. The use of nanoparticles could make pesticides more effective by reducing particle size to the nanoscale given the associated increase in surface area which introduces a fundamental change in the physicochemical properties of nano-pesticides [16, 17]. Compared to larger particles of the same chemical substance, nanoparticles are more reactive, more biologically active and have a more catalytic action [18, 19]. Nanoparticles could help use pesticides and fertilizers more effectively [16], for example, by reducing agrochemical components to nanosize or to pack the active ingredients in nanocapsules, which release them selectively, would allow for lower amounts with the same effect, only under certain conditions of heat, sunlight or pH [20, 21].

The current levels of application of nanoparticles and the expected developments to come, suggest that nanotechnology will have a direct impact on the evolution of pest management practices in agriculture [16, 22–25]. Recently, the discovery of nanoinsecticides brings new alternatives to expand the spectrum of applications of inorganic powders [6, 24, 26, 27]. Nanoengineered aluminium oxide as nanostructured alumina (NSA), has been shown to have insecticidal properties, low non target toxicity, non-reactivity, low cost and reduced probabilities of generating resistance in insects [26].

In a previous work, Stadler et al. [6, 24, 28] assumed that "*rice weevil*" (*S. oryzae*) adults acquire electrostatic charge by triboelectrification when walking on a dielectric surface such as wheat kernels, and that these charges on the insect body lead NSA particles from the treated substrate to the insect body surface. In order to verify this phenomenon empirically, studies were undertaken to examine and model the tribo-charging of *S. oryzae* adults on a dielectric surface and to identify the type and magnitude of the electrostatic charge on NSA, diatomaceous earth (DE) samples and the net electrostatic charge density of wheat kernels.

1.2. Nanostructured alumina: a novel nano-engineered insecticidal powder

A reduction in particle size of a substance results in increased surface/volume ratio per unit weight, which generally correlates with increased toxicity [29]. This characteristic has been exploited by some researchers to control various microorganisms and insects by applying nanoparticles [26, 28]. For example, nano-engineered alumina (NSA) is the result of combustion synthesis, using a redox mixture, with glycine as fuel and aluminum nitrate as oxidizer, where the final product is a homogeneous powder of high purity with uniform characteristics and specific physicochemical properties [30]. During the combustion process, alumina nanoparticles (40–60 nm) aggregate in primary clusters of approximately equal diameter building electrically loaded amorphous micrometric agglomerates/aggregates with a specific surface area of 14 m^2/g ranging in size from 0.1 to a few micrometers [31]. Due to its special characteristics as kinetics and bioactivity, it allows for varied and novel uses as a pesticide for human, livestock and agricultural use, stored product protection, treatments for wood preservation, carriers of pheromone or virus for insect pest control, etc. [32]. NSA has been shown to have strong insecticidal properties to several insect species through a mechanism of action different from conventional pesticides. Nanostructured alumina (NSA) has been shown to be an effective contact insecticide for several species of stored grain insect pests [6] as well as for leaf cutting ants [33].

In nanomaterials synthesized by oxidation of metals, such as the NSA, the resulting particles are electrically charged, showing a dipole-dipole interaction that promotes aggregate formation with resistance to dissociation forces [34]. Depending on the synthesis procedure, the greater part of the aggregates charged either positively or negatively and only some of these are dipoles [31]. Thus, the combustion manufacturing process is the main factor responsible for the affinity of particles with the triboelectrically charged body surface of different insect species (**Figure 1**) and as a consequence, also responsible for insecticidal activity. However, the morphology of nanoalumina agglomerates can be influenced by different variables during the synthesis such as substrate concentration, additives and calcination temperature which play a decisive role in the final morphology and characteristics of nanoalumina [35].

The effect of NSA on insects has been investigated through contact as well as dietary intake toxicity bioassays. Also, the *in vivo* toxicity and the *in vitro* cytotoxicity of NSA particles were screened as reviewed below.

1.3. Insecticidal effect of NSA on stored product insect pests

1.3.1. The contact toxicity of NSA against stored product insect pests

The contact toxicity of NSA was first investigated using dry powder applications at three different relative ambient humidity levels [6]. Tests were conducted simultaneously with enhanced diatomaceous earth, Protect-It®, to compare the efficacy of NSA to that of commercial insecticide powders. Two major stored grain pests *Rhyzopertha dominica* and *Sitophilus oryzae* were tested and significant delayed mortality was observed. Both species experienced significant mortality after 3 days of continuous exposure to treated wheat. Nine days after treatment, the median lethal doses (LD_{50}) observed ranged from 127 to 235 mg kg^{-1}. Results showed that NSA was more effective in killing *S. oryzae* than Protect-It® and was equally toxic to *R. dominica*. *R. dominica* was, less susceptible to inert powders than *S. oryzae* [32]. According to Subramanyam and Roesli [36], *S. oryzae* is among the most susceptible species

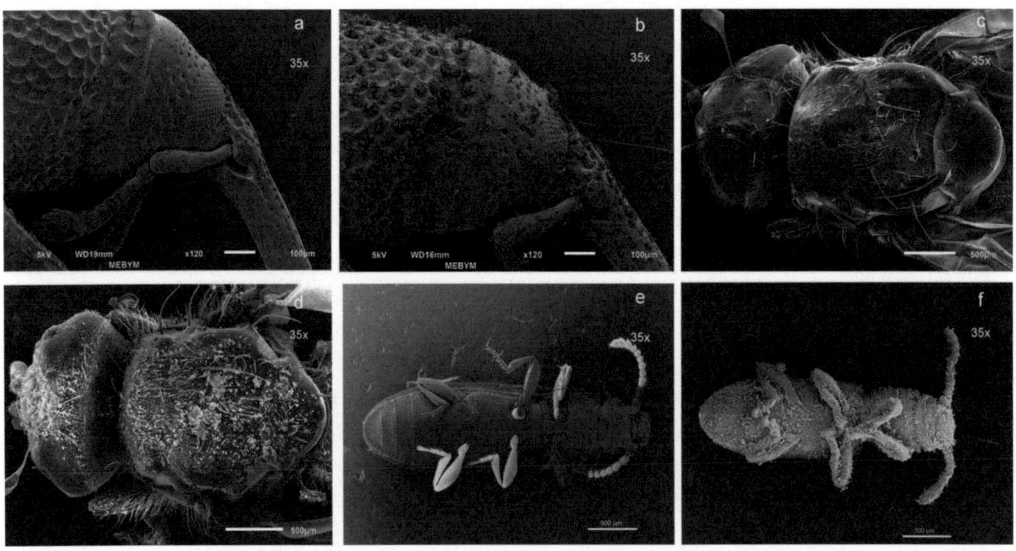

Figure 1. SEM images of individuals of three stored insect pest species [(a, b) *Sitophilus oryzae* (Coleoptera: Curculionidae); (c, d) *Ceratitis capitata* (Dipetara: Tephritidae); (e, f) *Oryzaephilus surinamensis* (Coleoptera: Silvanidae)] showing the affinity of NSA particles to triboelectrically charged insect body surfaces. (a, c, e) after exposure to untreated wheat kernels; (b, d, f) after exposure to 125 ppm NSA-treated wheat kernels. Fformat JEOL/EO, version 1.0; instrument JSM-6610.

to diatomaceous earth and *R. dominica* is among the least susceptible ones. Chemical makeup of epicuticular waxes varies across insect species [37], and this should translate into differences in susceptibility to nanoalumina and other inert powders due to differences in wetting. Treatment with NSA as well as Protect-It® also reduced progeny production although NSA powder was more effective in eliminating F1 adults than Protect-It®, for both species of insects tested. NSA reduced F1 progeny drastically at concentrations as low as 62.5 ppm for *S. oryzae* for high, medium and low humidity levels, and ranging from 250 to 500 ppm for *R. dominica* depending on the humidity level [32]. These results obtained with NSA are encouraging given that Protect-It® is one of the most effective DE-based products in the market [36, 38].

Comparison of these results with recommended rates for commercial insecticidal powders suggests that inorganic nanostructured alumina may prove a good alternative or complement to DE-based products, and encourage further testing with other insect pests and systems, plus experiments on delivery options to further enhance NSA products.

1.3.2. The intake toxicity of NSA in stored product insect pests

Although dehydration appears to be the main cause of mortality due NSA exposure, but it cannot be assumed as the only one, for it is found that at sub-lethal concentrations, insecticide powders in general exert further noxious effects on the insect [33, 36, 39]. Our studies revealed that intake toxicity is a significant mortality factor that occurs simultaneously with contact toxicity during insect exposure to NSA. Dietary intake of exposure to concentrations lower than 75 ppm caused sub-lethal effects in *S. oryzae* and adult mortality occurred after only 7 days exposure to NSA in food. These results indicate that toxicity due to ingestion is also a relevant mortality factor. There was a delayed response to NSA intake through ingestion which occurred up to 39 days of continuous exposure to NSA-treated flour discs. Mortality of adult *S. oryzae* was dose-dependent reaching up to 100% at concentrations of 250, 350 and 500 ppm, and up to 40% for concentrations below 125 ppm in wheat discs. The LC_{50} value calculated from intake bioassays on NSA-treated flour discs was 180.97 ppm [(CI = 167.07; 195.91); Slope = 0.01; Intercept = −1.68] and the LT_{50} calculated for the maximum dose concentration tested of 500 ppm was 23.82 days [(CI = 22.05; 25.17); Slope = 0.13; Intercept = −3.04]. The body weight of live individuals fed with NSA-treated wheat kernels (TPP plate No. 3), also presented a substantial reduction, of 51.6 (± 2.51)% on average [28]. These findings are similar to what Alexander et al. [40] observed after the treatment of *S. granarius* with various insecticide powders and by Trewin et al. [41] after treatment of *Ephestia kuehniella, Oryzaephilus surinamensis, Tenebrio molitor* and *Tribolium castaneum* with Aerosil® dispersions, a silica product. Results demonstrate that ingestion toxicity is a relevant long-term mortality factor that should be taken into account when assessing the efficacy of NSA and inert powders.

1.4. The *in vivo* toxicity and the *in vitro* cytotoxicity of NSA particles

An important characteristic of nanomaterials is their extremely large surface. For example, the same mass of material in the form of nanoparticles has a specific surface area which is many times larger than a coarse powder. This large surface can chemically react with materials that are otherwise non-reactive and non-toxic. However, it is not the size alone that contributes to the potential toxicity of nanomaterials. Rather, it has been shown that the toxicity of nanomaterials is influenced by many parameters like size, shape, surface area, electric charge and texture. Moreover, it is possible to design a basic nanomaterial in its morphology and surface properties on the nanotechnology. This results in new properties of a new material but it is unclear to what extent significant changes in toxicological properties can result from changes in the morphology [42]. Soluble nanomaterials lose their nanostructure features after they have come into contact with in biological fluids and a non-specific toxicity could possibly arise. However, insoluble nanomaterials that retain their nanostructures which are wrapped in a stable material and thus cannot come into contact with biological fluids or those that can be absorbed into organism are of lesser importance. Therefore, the free nanomaterials, which can be absorbed in different ways, are of particular toxicological significance. These nanomaterials are likely to lead to a higher exposure of humans and the environment [43].

Each nanostructured material has to be individually tested for its potential toxicity, since the knowledge about the complex relationships between physical and chemical parameters and a possible toxicity is missing. The toxicity of aluminum oxide nanoparticles has been discussed in many publications providing mixed results [44, 45]. On the other hand, toxicity of nanostructured aluminum oxide particles (NSA) [26] remains still an object of experimental work. Pochettino et al. [46] evaluated *in vitro* effect of the NSA on macrophages from the THP-1 cell line, exposed during two different time periods (6 and 24 hours) to different NSA concentrations (5, 25, 100 and 250 μg/mL). Cell cultures exposed to the lower concentrations of NSA during 6 hours show increased levels of the proinflammatory cytokine synthesized by macrophages IL-1β and a significant reduction of catalase (CAT) antioxidant enzyme activity. The two highest concentrations of NSA induced a decrease in cell viability (MTT assay) and an increase in lactate dehydrogenase activity (LDH: cytotoxicity indicator) and IL-1β release, in exposed cell cultures, and a decrease in CAT activity and thiol groups (–SH: thiols groups, antioxidants properties). These changes observed in CAT, LDH, –SH are indicators of oxidative stress. After NSA treatment, mitochondria lost their filamentous shape and displayed several morphological alterations. The effect of NSA on cell cultures after 24 hours of exposure was similar to that observed at 6 hours.

The exposure of THP-1 macrophage cell cultures to high NSA concentrations induces the release of IL-1β but also causes cell death, where NSA-mediated oxidative stress could play an important role. The generation of a controlled oxidative stress leads to the activation of intracellular mechanisms to compensate the production of reactive oxygen species (ROS), however a continuous overproduction of these species causes the onset of pathological states. Further studies should address the mechanisms involved in the oxidative stress caused by NSA in order to characterize and limit these. Also, further studies on the balance between pro- and anti-inflammatory molecules in *vitro* cell cultures exposed to NSA will be necessary looking for the mechanisms involved in acute effects of NSA exposure. On the other hand, low NSA concentrations raise the IL-1β levels without inducing changes in cell viability; so, this could be of relevance to enhance triggering immune responses. These results motivate further research on the mechanisms underlying the observed effects of NSA on THP-1 macrophage cells, as well as to analyze other mediators and immunological parameters in order to evaluate the potential of the NSA at low dose as a modulator of the immune response.

Deepening at the cellular level, Nadin [47] studied the genotoxicity effects of NSA at the cellular level. To determine whether NSA induces DNA damage, human peripheral blood mononuclear cells (PBL) were isolated from a healthy donor venous blood. PBL were exposed for 24 hours to increasing concentrations of NSA (50, 100 and 200 μg/mL) and then collected. Concentrations used were the same as those tested by Pochettino [46]. DNA and chromosomal damage was assessed throughout the alkaline comet assay and micronuclei (MIN) test, respectively, and cell viability was tested with the resazurin assay. The comet assay allowed to quantify DNA damage and revealed no significant increase in DNA damage induced by NSA. No statistical significant differences were found in terms of cellular viability and NSA had no significant effect on MIN induction.

Regarding animal experiments (*in vivo*), the acute oral toxicity and the acute inhalation toxicity of engineered aluminum oxide nanostructured particles (avg. 100 nm) were assessed in Wistar albino rats [48]. Acute oral toxicity was assessed by a limit test at a test dose of 2000 mg/kg b.wt that was administrated in a single dose. No mortality was observed in treated animals and no significant differences in body weight where observed ($p < 0.05$) either. No morphological changes where observed through pathological examinations. After inhalation exposure (0.02 mg/L air), respectively, during 4 hours, no changes in body weight gain were noted. A decrease in body weight gain was observed after inhalation exposure with 0.07 mg/L. No morbidity or mortality was observed in inhalation NSA exposed rats. These studies provide information applicable to the early stage in the hazard identification process for this type of nanomaterials that could be useful in risk management in the context of production, handling and use of nanomaterials. These results show that acute oral and inhalation exposure to NSA did not result in morbidity or mortality in male rats.

The rapid proliferation of engineered nanomaterials and the limited toxicological data currently available on it presents a dilemma to regulators regarding risk assessment processes for these materials [49]. For recently developed nanomaterials, there are in many cases insufficient investigations into health effects. Therefore, no sufficiently reliable statement can be made about these nanomaterials. There is a need to determine the extent of absorption, systemic availability, accumulation and excretion of nanomaterials after inhalation and oral exposure. However, the necessary in vivo studies should be integrated into any toxicological studies to avoid unnecessary animal experiments. The influence of modifications in the NSA synthesis on the kinetic parameters [26] as well as on the toxicological properties of the nanomaterials should also be examined. Finally, oxidative stress and the formation of reactive oxygen species (ROS) are fundamental key mechanisms of cellular defense after particle capture.

In order to develop biorational pesticides through design of NSA synthesis, further research is necessary on the complex relationships between its physical and chemical parameters and its toxicity.

2. Identifying and understanding the mechanism of pesticide action of nanostructured alumina

2.1. Triboelectric charging in insects

Tribo-charging is the advent of electric charges based on the mechanisms of charge transfer which occurs when two different non-conductive bodies (materials) are brought into contact and separated or rubbed together acquiring positive or negative polarity [50]. Friction plays only a role in this respect, as the bodies are approached to molecular distances, thus permitting charge transfer (contact electricity). A triboelectric series can be established for the frictional electrification in which a material is positively charged when friction is applied to the following material, while friction is negative in the previous one. This series is based on Cohen's rule according to which the substance with the higher dielectric constant is positively charged [51].

Insects also generate electrostatic charges by walking. This was first studied by Edwards [52, 53] who showed that rubbing dead insects against various substrates generated electrostatic charges. In a later study [54], this author monitored naturally acquired and retained electrostatic charges on living insects, showing that a net charge could be detected in flying insects. For example, a flying honeybee in a wind-tunnel reaches an average charge of −23.1 pC [55] and this charge plays a key in the transfer of pollen grains from the flower to the insect [56]. Corbet et al. [57] showed that due to electrostatic charges, oilseed rape pollen grains pass from flower to freshly killed honeybee across an air gap of 0.5 mm. Electrostatic charges in insects may arise from frictional charging linked to contact with different types of surfaces through the migration of electrons from one surface to another, where equal but opposite charges arise on each surface [58–60]. However, insects may also acquire electrostatic charge by absorption via the insect cuticle through dermal pores [61], as well as through the adhesion of charged particles [55, 62, 63].

2.2. Electrostatic charge of insecticidal powder particles

Powders or more generically, solids in a high degree of subdivision, exist in an enormous variety of chemistries and morphologies. The discrete entities or "kinetic units" of interest typically range in linear dimension from a few micrometers to a few nanometers, at the colloid size range. Even in the nano-range, where powders take the form of quantum dots or nanowires, the objects are amenable to the descriptions afforded by macroscopic thermodynamics [64]. These particles tend to sediment from the air due to their greater density, depending on the environmental conditions and the shape and size of the particles. In dry atmospheres, the sedimentation or sink rate of the particles can be calculated as a function of their radius [65]. After landing, an adhesion process occurs immediately after the particle hits the surface and is a purely physical process. It is relatively weak, reversible and is based on unspecific capillary, van der Waals, electrostatic and hydrophobic forces between the particle and the surface [66]. These forces have a different strength and they also differ in their range. In order to get into the area of influence of molecular interactions, two surfaces have to approach below 10 nm. Capillary forces act in a range of 10–200 nm and electrostatic forces of 100nm–1 μm [67]. In some studies, it was found that there is an influence of surface hydrophobicity on adhesion [68, 69]. Thus, a stronger adhesion of particles to hydrophobic than to hydrophilic surfaces was detected. Furthermore, it has been shown that the surface roughness also has an influence on adhesion of the particles [67].

2.3. Electrostatic charge of wheat kernels

General characteristics of wheat seeds depend on a wide range of dielectric properties like conductance and bioelectric potentials related to ionic and structural heterogeneity of plant cells, tissues and organs. Biologically active substances as enzymes, contribute to bioelectric polarity through powerful charge at the molecular level [70]. However, ionic activities inside the tissues dominate the low-frequency dielectric behavior of the tissues [71]. Additionally, the structure and shape of epidermal cells and epicuticular waxes of wheat seeds also contribute to their bioelectric activity [72]. Nonetheless, the bioelectric activity of a plant is an intrinsic structural feature of the organism and cannot be modified since it is genetically predetermined [70].

Bioelectric polarity is critical to adhesion or repellence of water or particles of different nature, shape and size from any surface [73, 74]. When particles, of whatever nature, reach a surface as, for example, the wheat seed epidermis (*testa*), interactions occur between particles and surface. If the particles are in the range of millimeters or above, gravitation and mass inertia are the decisive forces for these interactions where adhesion forces dominate [75]. These forces consists of different forces as capillary force, electrostatic force (Coulomb repulsion/attraction of different surplus charges, electrostatic double layer force) and molecular interactions (van der Waals forces, dipole-dipole interactions and hydrogen bonds) [76].

3. Assessment of tribo-charging in insects, electrostatic charge of insecticidal powder particles and wheat kernels

3.1. Materials and methods

3.1.1. Insects

Sitophilus oryzae (Linnaeus, 1763) (Insecta, Coleoptera: Curculionidae) were obtained from the Laboratory of Environmental Toxicology (IMBECU.CONICET, Argentina) culture, reared on wheat kernels (var. Baguette NIDERA) at 27 ± 2°C, 70 ± 5% RH in the dark. Adults used in all experiments were of unknown sex, mating status and age.

3.1.2. Insecticide powders

3.1.2.1. Nanostructured alumina (NSA)

Synthesized since Toniolo et al. [30] by glycine-nitrate combustion technique using a redox mixture, with glycine as fuel and aluminum nitrate nonahydrate as oxidizer. Nanostructured particles sized from approximately 0.1 μm up to a few micrometers.

3.1.2.2. Diatomaceous earth

Commercial diatomaceous earth (DiatomiD®) from fossilized sedimentary phytoplankton microalgae (diatoms) deposits from San Juan-Argentina, which contains over 85% amorphous SiO_2 and particles sizing from 1 to about 150 μm.

3.2. Experimental setup

Triboelectric charges on insects as well as charge densities on wheat kernels and insecticide powders were assessed under the same experimental and environmental conditions by means of a Faraday cup connected to an ensemble of an electrometer based on a LMP7721 amplifier (NI, LMP7721 Multi-Function Evaluation Board amplifier in buffer mode) and a data acquisition system (NI USB 6009 (8 input, 14 bits, multifunction I/O, 10 bits DAQ system) controlled by NI Labview software (EFC). The detection limit of the EFC was 0.06pC. Electrometer calibrations were performed using ADA4530-1R-EBZ-BUF as the reference electrometer. Total electrometer input capacitance was assessed with Analog Devices AN-1373. The tribo-charging assessment method was validated since Greason [77] by using a stainless steel sphere (ø 2 mm) sliding along a slightly inward curved paperboard ramp (length 400 mm and 50 mm wide) coated with a smooth layer (1.5 ± 0.5 mm) of dried wheat paste (wheat flour and water). The ramp was tilted at 30°, so the stainless steel sphere slides into the Faraday cup at the end of the ramp.

Experiments were conducted within a grounded Faraday cage to avoid external sources of static electricity. In order to set the same baseline for each experiment, grounding was used to neutralize the initial charges carried by samples. Throughout data collection, the operator remained connected to the grounded Faraday cage. Temperature and humidity inside the Faraday cage were maintained at 25° ± 2°C and 35 ± 5% RH and were constantly monitored during experiments.

Tribo-charging in *S. oryzae* was measured by using the paperboard ramp and EFC. Frictional charging experiments were developed by using live CO_2 anesthetized *S. oryzae* adults sliding smoothly from different distances on the ramp (1.25, 2.50, 5.00, 7.50, 10.0, 12.5, 15.0, 20.0, 25.0, 30.0 and 40.0 cm). The insects slid at an almost constant speed under the action of gravity and fell into the Faraday cup down to the end of the ramp. The charge on the insect was detected by the EFC and the data were automatically stored in a computer. The process was repeated 12 times for each distance using different insects.

3.2.1. *Assessment of electrostatic charge on insecticide powders*

Charge density of nanostructured alumina (NSA) synthesized since Toniolo et al. [30] and diatomaceous earth (DE) [DiatomiD®] was measured by the static method [78]. Identical volumes of the inert powders were measured at 25°C, 35% RH, using a normalized copper cylinder (h = 3.2 mm; r = 8.75 mm, internal). By means of the earthed 0.769 mL cylinder, samples of 0.23 g of nanostructured alumina and on the other hand 0.74 g of diatomaceous earth were transferred into the Faraday cup. The process was repeated 20 times using always the same insecticide powder samples.

3.2.2. *Assessment of electrostatic charge on wheat kernels*

Electrostatic charge density of seed was measured by distributing 20 selected wheat kernels (55.2 mg/kernel (SD ±8.8 10^{-3}) var. Baguette NIDERA (4 months after harvest) in a single layer on a grounded copper plate. Six randomly selected kernels were introduced one at a time, for 12 times each in the Faraday cup (EFC) under the experiments conditions described above.

http://dx.doi.org/10.5772/intechopen.72448

4. Results

4.1. Tribo-charging in insects

Figure 2 shows tribo-charging of *S. oryzae* where the rate of charging at the start was proportional to the saturation charge and it decreased as the insects charge increased. The insect loses electrons as far as maximum charge is attained when the electron affinities reach equilibrium. The charge on the ramp surface has no influence on its particular electron affinity since the insect in motion rub sequentially different and uncharged sections of the ramp surface during sliding. The charge acquired by the insect with each additional distance covered on the ramp is equivalent to the difference between the insect maximum reachable charge and the charge of the ramp surface [58].

As shown in **Figure 2**, the magnitude of electric charge picked up by *S. oryzae* was approximately proportional to the distance it moved ($d_{1.25cm}$ = +0.766 (±0.254) pC/insect to $d_{40.0cm}$ = +2.560 (±0.221) pC/insect). In contrast to McGonigle et al. [58] and in some extent in concordance with Jackson and McGonigle [60], our results show a discrete evidence for a plateauing of charge and clearly demonstrate that saturation charge in *S. oryzae* was not reached (**Figure 2**).

4.2. Electrostatic charge in insecticide powders

The magnitude and sign of the net average electrostatic charge density measured was −93.91 (±2.62) pC/grain for NSA and −11.554 (±2.342) pC/grain for diatomaceous earth. Thus, both substances are negatively charged and consequently adhere on electropositive insects body surfaces.

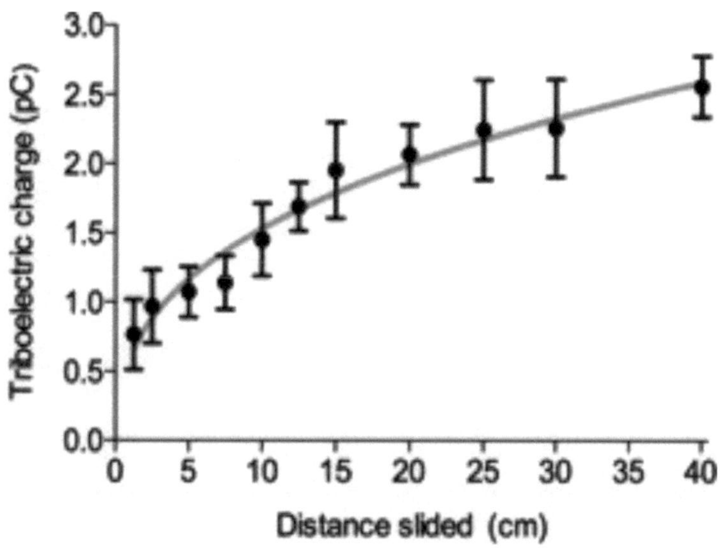

Figure 2. Mean charge (pC) generated by live anesthetized *S. oryzae* adults after sliding along different wheat flour ramp track sections (1.25–40 cm). Experimental were plotted (dots) alongside a modeled curve (entire line).

4.3. Electrostatic charge in wheat kernels

The electrostatic charge measured on wheat kernels var. Baguette NIDERA was weakly negative, averaging −0.191 (± -7.15×10^{-2}) pC/grain.

5. Discussion

In principle, all life forms are immersed in an ionized environment. Ions bear an electric charge and thus an electric field may be influenced by another one. Thus, the electric fields from two bodies would interact in such a way that initially the ions would be driven or set into motion.

Our experiments showed that electrostatic charge in wheat seeds is weakly negative (−0.191 (±-7.15×10^{-2}) pC/grain), the electrostatic charge of diatomaceous earth is slightly negative (−11.554 (±2.342) pC/grain) and nanostructured alumina bears a strong negative electrostatic charge (−93.91 (±2.62) pC/grain). These data indicate despite the negative charge of wheat kernels, other characteristics such as rugosity and hairs on wheat kernels' surface are determinant for the surface attachment of DE and NSA particles (**Figure 3a, b**). Thus, even the repelling force between like charged particles and wheat kernels, the low net charge density of these will not be relevant enough for particle detachment from the kernels and therefore, the smaller the particles the denser the wheat grain surface coverage (**Figure 3b**).

As shown here and by different authors [55, 58, 60], insects possess bodily electric charges raised by walking or flying. In our experiments, the insects rubbing against flour on the 30° tilted ramp emulate their movement within a stored grain matrix where they charge themselves throughout friction (tribo-charging) and thereby enhance adherence of all particles bearing an opposite charge to their body.

As shown, the rate of insect tribo-charging at the start of the ramp was proportional to the saturation charge that decreases as the insects charge increases. This can be explained as follows: a sliding insect can be thought of as a conducting but electrically isolated object in motional contact with the ramp. The insect and ramp surface start with unequal electron affinities. The ramp surface has a high electron affinity so it takes electrons from the insect gaining negative charge and the insect gains a positive charge due to the loss of electrons [79].

The experimental results presented here (**Figure 2**) show that adults *S. oryzae* take up and retain a positive electrostatic charge on the cuticle, approximately proportional to the distance shifted on the experimental wheat flour ramp ($d_{1.25cm}$ = +0.766 (±0.254) pC/insect to $d_{40.0cm}$ = +2.56 (±0.221) pC/insect), which is consistent with the results obtained by Jackson and McGonigle [60] experiments. Thus, when *S. oryzae* was exposed to wheat kernels treated with NSA and/or DE dry powder, negatively charged particles became attracted to the positive tribo-charged insect body surface. However, bonding of DE particles on the insect body surface is 8.13 times weaker than NSA due to lower electric net charge (−11.554 (±2.342) pC/grain) of DE and its larger particle size (**Figure 4b**) and mass. Instead, bonding of NSA particles to the insect body surface is strong due the magnitude of its electric charge (−93.91 (±2.62) pC/grain) and because particles are smaller and lighter (**Figure 4a**).

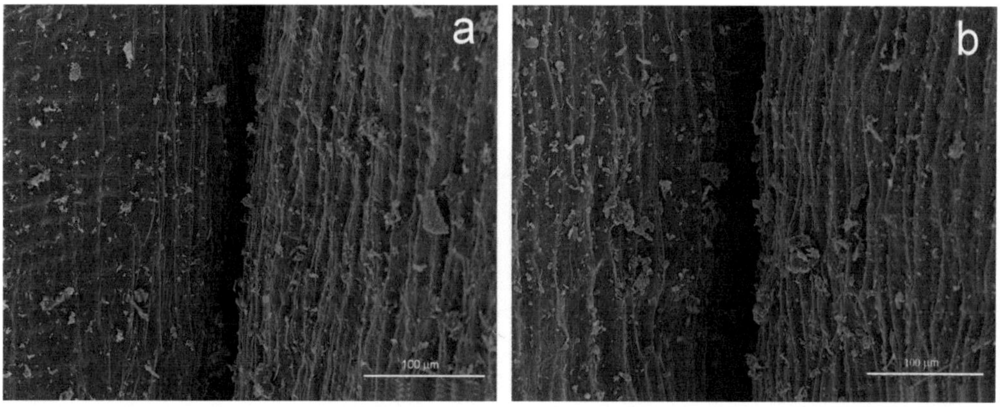

Figure 3. SEM image (300×) of a wheat grain *Triticum sativum* var. Baguette NIDERA. a – exposed to 125 ppm DE; b – exposed to 125 ppm NSA. Format JEOL/EO, Version 1.0; Instrument JSM-6610.

These differences in attachment effectiveness are evidenced by the fact that insects exposed to surfaces treated with NSA became massively and uniformly coated with NSA particles (**Figure 5a**). In contrast, insects exposed to surfaces treated with DE showed a scant and diffuse distribution of particles on their body surface (**Figure 5b**) demonstrating that DE are not retained as NSA particles are (**Figure 5c**).

Insecticidal inert powders in general, attach to the insect cuticle surface (**Figure 1**) damaging the cuticle and producing a negative effect on insect water balance [36]; furthermore insecticidal efficacy decreases as ambient humidity increases and this may negatively impact the efficacy of inorganic powder insecticides [26, 32]. This decrease in efficacy in at higher relative ambient humidity of abrasive powders as DE can be explained by a delayed drying process [80] due to a slower rate of water loss through the damaged insect cuticle [40, 81–84]. The natural transpiration rate of an insect into the surrounding air is dependent on water vapor pressure. With increasing relative humidity the vapor pressure increases in the air and the water discharge from the insect body surface tends to decrease. These results are consistent with earlier findings for abrasive insecticide powders which suggested that toxicity of insecticide powders on arthropods is a consequence of the "cuticular water flux" [84]. On the other hand, the loss of insecticide efficacy in sorbtive insecticide powders such as NSA at higher relative humidity can be explained by analyzing the effect of moisture on the interaction of tribocharged insect body surface and the small but high electrically charged particles of NSA as follows: at constant temperature, a substance absorbs moisture from the air until the material and humidity are in equilibrium attaining the adsorption isotherm of the substance [85]. As shown, triboelectric charge is the main reason for insecticide powder adhesion to insect body surface. In general, the importance of the triboelectric effect increases with low humidities and with smaller particles. High relative humidity can influence the interparticle forces when certain quantity of water is condensed on particle surface reducing the electrostatic forces by electrostatic discharge [86]. Initially, at lower humidity (<65%) [87], the water adsorbs on the particle in the form of water vapor. So, the interparticle bond forces can be reduced as electrostatic forces are reduced. As the humidity increases exceeding the critical value, capillary condensation occurs at the contact points of the particles and liquid bridges form. Above

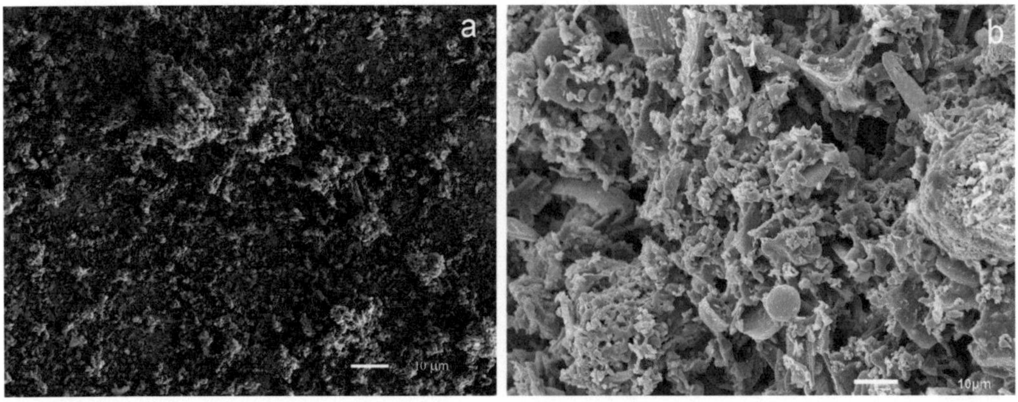

Figure 4. Scanning electron microscopy image of: a – nanostructured alumina (NSA) particles; b – diatomaceous earth (DiatomiD®). Format JEOL/EO, Version 1.0. Instrument JSM-661; AccelVolt 10. Mag 400; Signal SEI; Spot_Size 35. Vac Mode HV.

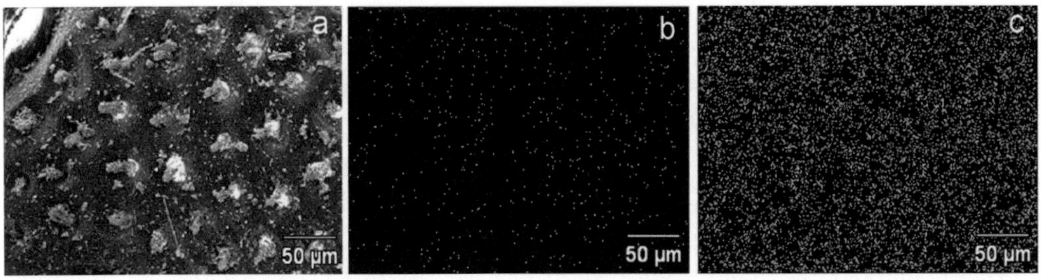

Figure 5. Prosternum of *S. oryzae*: a – SEM image of *S. oryzae* exposed to untreated wheat kernels (400×); b – exposed to wheat kernels treated with 125 ppm DE, silicon (*Si*) counts from Energy Dispersive Spectroscopy (EDS); c – exposed to wheat kernels treated with 125 ppm NSA, aluminum (*Al*) counts from EDS. Format JEOL/EO, Version 1.0; Instrument JSM-6610; Acc. Volt 10; Mag 400; Spot Size 35; Vac Mode HV. c – Aluminum (*Al*) counts from Energy Dispersive Spectroscopy (EDS) – Filter Fit $\chi 2$ value: 31.161; Errors: ±2; Sigma Correction Method: Proza (Phi-Rho-Z); Acc. Voltage: 12.0 kV; Take off angle: 35.9°.

the critical value, the capillary forces are the predominant forces [87]. Due to the different contact angles, hydrophilic substances as DE [88, 89] are more exposed to the influence of moisture than hydrophobic materials (synthesized Al_2O_3; [90]). In addition, water adsorption also affects the surface energy of the particles [91]. Similar to liquids, solids have an imbalance in the surface forces. However, in solids the molecules are much more strongly bonded to one another and the surface energy is not evenly distributed on the particle surface.

Differences in insecticidal efficacy between DE an NSA arise from structural and physical differences between these two products. DE combines high abrasive and low sorbtive properties due to sharp angular structure and large particle size (1 to about 150 μm [92]), (**Figure 3b**) and a relatively low specific surface area (ca. 4 m^2/g) [93]. In contrast, NSA particles (**Figure 3a**) are small aggregates (≈1.5 μm; [32]) assembled by coarse accumulations of nanoparticles (40–60 nm) which increase the overall specific surface area of the powder (ca.14 m^2/g [94]). Thus, DE insecticidal efficacy is lower than that of NSA due to its small electric affinity to the insect body surface (**Figure 4b**)

http://dx.doi.org/10.5772/intechopen.72448

in addition to low sorbtive properties. So, DE works, in general, stochastically by damaging the insect body surface mechanically when it moves within a stored grain matrix. On the other hand, NSA's high insecticidal efficacy depends on its increased electrical affinity to the insect body surface (**Figure 4c**) in addition to having greater sorbtive properties. The whole mechanism of action consists of two steps in sequential order. First, there is a strong electrical binding between negatively charged NSA particles and the positive tribo-charged insect. Next, dehydration of the insect occurs due to strong sorptive action of NSA particles removing the insect cuticular waxes responsible for protecting insects against water loss. Hence, the mechanism of action of NSA does target the water balance of the insect and dehydration is the leading cause of death.

6. Conclusions

Nanostructured alumina (NSA) is a nano-engineered material which has insecticide properties. The current study investigated its mode of action and demonstrated that tribo-charging is a key aspect in the interaction of NSA and the insects' cuticle. In fact, triboelectric charge is the main reason for insecticide powder adhesion to the insect body surface, and could explain at least in part, the efficacy differences observed in previous studies between NSA and diatomaceous earth (DE). Insects exposed to surfaces treated with NSA became massively and uniformly coated with NSA particles while insects exposed to surfaces treated with DE showed a scant and diffuse distribution of particles on their body surface. This in turn, was accompanied by a difference in charge between both powders, where NSA has a greater intrinsic electric charge than DE. Moreover, NSA charges did not decay as a consequence of NSA low wettability. Thus, the current study supports previous studies showing that NSA has a greater affinity towards the insect cuticle and a greater insecticidal efficacy than other inert powders, and provides a reasonable explanation of its mechanism of action through triboelectric and sorbtive phenomena. Further research is necessary to contribute to the knowledge of the complex relationships between physical and chemical parameters of insects and powders, responsible for insecticide activity. Future studies should focus on determining the insect chemical and physical characteristics that are involved in toxicity of inert powders such as NSA to insects. Measuring the tribolectric charges of different insect species could shed light on the basis of these differences in toxicity observed among different insect species to NSA, which may be related to their chemical composition as well as their physical structure, leading to electric charges of different sign and magnitude.

With regards to toxicity research studies should aim to determine the extent of absorption, systemic availability, accumulation and excretion of nanomaterials after inhalation and oral exposure, as well as genotoxicity. However, the necessary in vitro studies should be integrated into any toxicological studies to avoid unnecessary animal experiments. The influence of modifications in the NSA synthesis on the kinetic parameters as well as on the toxicological properties of the nanomaterials should also be examined. Finally, oxidative stress and the formation of reactive oxygen species (ROS) are fundamental key mechanisms of cellular defense after particle capture.

The current study, investigating the mode of action of NSA, supports previous studies demonstrating that NSA is more effective than other insecticide powders and has good potential as insecticide of stored grain insect pests since it possesses some of the characteristics of an ideal insecticide, given that it is not reactive, of low synthesis cost, with reduced probabilities of generating resistance in insects, and it is more effective than other commercially available insecticidal powders. It is likely that NSA may be used against other insect pests with similar and further research investigating this is warranted.

Acknowledgements

The authors would like to thank Lic. María S. Lassa (MEByM-CONICET) for providing technical assistance in SEM imaging and EDS analysis and to Mr. Daniel E. Dueñas (CONICET, CCT-Mendoza) for high quality editing service. We are also grateful to Ing. Ramiro Oviedo Bustos (CUME SRL) and Ing. Mauro Mortarini (Ojos del Salado SRL) for contributing to our research, as well as to Ing. Mariano A. Bernardo (NIDERA SA) for providing standard seed material for bioassays. This work was supported by grant PIP No. 11220130100382CO from the National Scientific and Technical Research Council, Argentina (CONICET).

Author details

Teodoro Stadler[1*], Micaela Buteler[2], Susana R. Valdez[3,4] and Javier G. Gitto[1,5]

*Address all correspondence to: lpe@mendoza-conicet.gob.ar

1 Laboratorio de Toxicología Ambiental, Instituto de Medicina y Biología Experimental de Cuyo (IMBECU), CONICET, Centro Científico Tecnológico-Mendoza, Mendoza, Argentina

2 INIBIOMA, CONICET – Universidad Nacional del Comahue, Bariloche, Argentina

3 Laboratorio de Reproducción y Lactancia, Instituto de Medicina y Biología Experimental de Cuyo (IMBECU), CCT – CONICET, Mendoza, Argentina

4 Facultad de Ciencias Exactas y Naturales, Universidad Nacional de Cuyo, Mendoza, Argentina

5 Laboratorio de Electrostática y de Materiales Dieléctricos, Universidad Tecnológica Nacional, Facultad Regional Mendoza, Argentina

References

[1] Aktar MW, Sengupta D, Chowdhury A. Impact of pesticides use in agriculture: Their benefits and hazards. Interdisciplinary Toxicology. 2009;**2**(1):1-12. DOI: 10.2478/v10102-009-0001-7

[2] Beketov MA, Kefford BJ, Schafer RB, Liess M. Pesticides reduce regional biodiversity of stream invertebrates. Proceedings of the National Academy of Sciences USA. 2013;**110**:11039-11043. DOI: 10.1073/pnas.1305618110

[3] Gill HK, Garg H. Pesticides: Environmental Impacts and Management Strategies. Croatia: InTech; 2014. DOI: 10.5772/57399. Available from: https://www.intechopen.com/books/pesticides-toxic-aspects/pesticides-environmental-impacts-and-management-strategies

[4] Pretty J, Bharucha ZP. Integrated pest management for sustainable intensification of agriculture in Asia and Africa. Insects. 2015;**6**(1):152-182. DOI: 10.3390/insects6010152

[5] Duke SO, Cantrell CL, Meepagala KM, Wedge DE, Tabanca N, Schrader KK. Natural toxins for use in Pest management. Toxins. 2010;**2**:1943-1962

[6] Stadler T, Buteler M, Weaver DK. Novel use of nanostructured alumina as an insecticide. Pest Management Science. 2010;**66**(6):577-579. DOI: 10.1002/ps.1915

[7] Sparks TC, Nauen R. IRAC: Mode of action classification and insecticide resistance management. Pesticide Biochemistry and Physiology. 2015;**12**:122-128. DOI: doi.org/10.1016/j.pestbp.2014.11.014

[8] Issa B, Obaidat IM, Albiss BA, Haik Y. Magnetic nanoparticles: Surface effects and properties related to biomedicine applications. International Journal of Molecular Sciences. 2013;**14**(11):21266-21305. DOI: 10.3390/ijms141121266

[9] Kennedy D. The role and future of nanotechnology in research, industry and education. In: MATRIB '06: Conference on Materials, Processes, Friction and Wear; Vela Luka. 22-24 Jun, 2006. 2006. Available from: http://arrow.dit.ie/cgi/viewcontent.cgi?article=1038&context=engschmeccon

[10] Somorjai GA, Li Y. Impact of surface chemistry. Proceedings of the National Academy of Sciences of the United States of America. 2010;**108**(3):917-924. DOI: 10.1073/pnas.1006669107

[11] Falagan-Lotsch P, Grzincic EM, Murphy CJ. New advances in nanotechnology-based diagnosis and therapeutics for breast cancer: An assessment of active-targeting inorganic nanoplatforms. Bioconjugate Chemistry. 2017;**28**(1):135-152. DOI: 10.1021/acs.bioconjchem.6b00591

[12] Miller C, Sarewitz D, Light A. Science, technology, and sustainability: Building a research agenda. National Science Foundation Supported Workshop. 2008. Available from: https://www.nsf.gov/sbe/ses/sts/Science_Technology_and_Sustainability_Workshop_Rpt.pdf

[13] OECD. OECD Green Growth Studies Policy Instruments to Support Green Growth in Agriculture. France: OECD Publishing; 14 Oct 2013. 140 p. DOI: http://dx.doi.org/10.1787/9789264203525-en

[14] Kah M, Hofmann T. Nanopesticide research: Current trends and future priorities. Environment International. 2014;**63**:224-235. DOI: https://doi.org/10.1016/j.envint.2013.11.015

[15] Jahanban L, Reza Davari M. Building organic bridges. Organic World Congress 2014, Istanbul, Turkey. Proceedings of the 4th ISOFAR Scientific Conference. 13-15 Oct 2014. Eprint ID 23620

[16] Sekhon BS. Nanotechnology in agri-food production: An overview. Nanotechnology, Science and Applications. 2014;**7**:31-53. DOI: 10.2147/NSA.S39406

[17] Khan MR, Rizvi TF. Nanotechnology: Scope and application in plant disease management. Plant Pathology Journal. 2014;**13**:214-231. DOI: 10.3923/ppj.2014.214.231

[18] Mukhopadhyay S. Nanotechnology in agriculture: Prospects and constraints. Nanotechnology, Science and Applications. 2014;**7**:63-71. DOI: 10.2147/NSA.S39409

[19] Navya PN, Daima HK. Rational engineering of physicochemical properties of nanomaterials for biomedical applications with nanotoxicological perspectives. Nano Convergence. 2016;**3**(1):1-14. DOI: 10.1186/s40580-016-0064-z

[20] Sopeña F, Cabrera A, Maqueda C, Morillo E. Controlled release of the herbicide norflurazon into water from ethylcellulose formulations. Journal of Agricultural and Food Chemistry. 2005;**53**(9):3540-3547. DOI: 10.1021/jf048007d

[21] Werdin González JO, Gutiérrez MM, Ferrero AA, Band BF. Essential oils nanoformulations for stored-product pest control – Characterization and biological properties. Chemosphere. 2005;**100**:130-138. DOI: 10.1016/j.chemosphere.2013.11.056

[22] Rai M, Ingle A. Role of nanotechnology in agriculture with special reference to management of insect pests. Applied Microbiology and Biotechnology. 2012;**94**:287-293. DOI: 10.1007/s00253-012-3969-4

[23] Pramanik S, Pramanik G. Nanotechnology for sustainable agriculture in India. In: Ranjan S, Dasgupta N, Lichtfouse E, editors. Nanoscience in Food and Agriculture. 3rd ed. Vol. 23. Sustainable Agriculture Reviews. Cham: Springer; 2016. pp. 243-280. DOI: https://doi.org/10.1007/978-3-319-48009-1_10

[24] Stadler T, Buteler M, Weaver DK. Nanoinsecticidas: Nuevas perspectivas para el control de plagas. Revista de la Sociedad Entomológica Argentina. 2010;**69**(3-4):149-156 Available from: http://www.scielo.org.ar/pdf/rsea/v69n3-4/v69n3-4a01.pdf

[25] Sandhya M, Chetan K, Abhilash PC, Fraceto LF, Singh HB. Integrated approach of agri-nanotechnology: Challenges and future trends. Frontiers in Plant Science. 2017;**8**:1-12. DOI: https://doi.org/10.3389/fpls.2017.00471

[26] Buteler M, Sofie SW, Weaver DK, Driscoll D, Muretta J, Stadler T. Development of nanoaluminadust as insecticide against *Sitophilus oryzae* and *Rhyzopertha dominica*. International Journal of Pest Management. 2015;**6**:80-89. DOI: http://dx.doi.org/10.1080/09670874.2014.1001008

[27] Debnath N, Das S, Seth D, Chandra R, Bhattacharya SC, Goswami A. Entomotoxic effect of silica nanoparticles against *Sitophilus oryzae* (L.). Journal of Pest Science. 2011;**84**:99-105. DOI: https://doi.org/10.1007/s10340-010-0332-3

[28] Stadler T, López García GP, Gitto JG, Buteler M. Nanostructured alumina: Biocidal properties and mechanism of action of a novel insecticide powder. Bulletin of Insectology.

2017;**70**(1):17-25 Available from: http://www.bulletinofinsectology.org/pdfarticles/vol70-2017-017-025stadler.pdf

[29] Paull J, Lyons K. Nanotechnology: The next challenge for organics. Journal of Organic Systems. 2008;**3**:3-22 Available from: http://orgprints.org/13569/1/13569.pdf

[30] Toniolo JC, Lima MD, Takimi AS, Bergmann CP. Synthesis of alumina powders by the glycine-nitrate combustion process. Materials Research Bulletin. 2005;**40**:561-571. DOI: 10.1016/j.materresbull.2004.07.019

[31] Karasev VV, Onischuk AA, Glotov OG, Baklanov AM, Maryasov AG, Zarko VE, Panfilova VN, Levykin AI, Sabelfeld KK. Formation of charged aggregates of Al_2O_3 nanoparticles by combustion of aluminum droplets in air. Combustion and Flame. 2004;**138**:40-54. DOI: 10.1016/j.combustflame.2004.04.001

[32] Stadler T, Buteler M, Weaver DK, Sofie S. Comparative toxicity of nanostructured alumina and a commercial inert dust for *Sitophilus oryzae* (L.) and *Rhyzopertha dominica* (F.) at varying ambient humidity levels. Journal of Stored Products Research. 2012;**48**:81-90. DOI: 10.1016/j.jspr.2011.09.004

[33] Buteler M, Lopez Garcia GP, Stadler T. Potential of nanostructured alumina for leaf-cutting ants *Acromyrmex lobicornis* (Hymenoptera: Formicidae) management. 2017. Version of Record Online: 23 Mar 2017. DOI: 10.1111/aen.12277

[34] Chacon RR. 2007. La alúmina como material aislante en la fusión termonuclear. Efecto de la incorporación de carbono en las propiedades físicas. [thesis]. Universidad Carlos III de Madrid: Departamento de Ciencia de los Materiales e Ingeniería Metalúrgica Instituto Álvaro Alonso Barba; 2007

[35] Xie Y, Kocaefe D, Kocaefe Y, Cheng J, Liu W. The effect of novel synthetic methods and parameters control on morphology of Nano-alumina particles. Nanoscale Research Letters. 2016;**11**:259. DOI: https://doi.org/10.1186/s11671-016-1472

[36] Subramanyam B, Roesli R. Inert dusts. In: Alternatives to Pesticides in Stored-product IPM. US: Springer; 2000. pp. 321-380. DOI: https://doi.org/10.1007/978-1-4615-4353-4_12

[37] Nawrot J, Maliński E, Szafranek J. Function and composition of cuticular hydrocarbons of stored-product insects. In: Highley E, Wright EJ, Banks HJ, Champ BR, editors. Stored Product Protection. Proceedings of the 6th International Working Conference on Stored-Product Protection; 17-23 Apr 1994; Canberra, Australia. Wallingford (UK): CAB International; 1994

[38] Korunic Z, Fields P. U.S. Patent No. 5773017. Washington, DC: U.S. Patent and Trademark Office; 1998

[39] Kabir BGJ, Lawan M, Gambo FM. Efficacy and persistence of raw diatomaceous earth against *Tribolium castaneum* Herbst (Coleoptera: Tenebrionidae) on stored maize, sorghum and wheat. Academic Journal of Entomology. 2011;**4**:51-58 Available from: https://www.idosi.org/aje/4(2)11/4.pdf

[40] Alexander P, Kitchener JA, Briscoe HVA. Inert dust insecticides: Part II. Mechanism of action. Annals of Applied Biology. 1944;**3**:150-156. DOI: 10.1111/j.1744-7348.1944.tb06226.x

[41] Trewin B, Reichmuth C. Wirksamkeit des Kieselgurpräparates Dryacide® gegen vorratsschädliche Insekten. Anzeiger für Schädlingskunde, Pflanzenschutz, Umweltschutz. 1997;**70**(3):51-54. DOI: https://doi.org/10.1007/BF01996921

[42] Wyser Y, Adams M, Avella M, Carlander D, Garcia L, Pieper G, Rennen M, Schuermans J, Weiss J. Outlook and challenges of nanotechnologies for food packaging. Packing Technology and Science. 2016;**29**(12):615-648. DOI: 10.1002/pts.2221

[43] Dowling AP. Development of nanotechnologies. Materials Today. 2004;**7**(12):30-35. DOI: https://doi.org/10.1016/S1369-7021(04)00628-5

[44] Ray PC, Yu H, Fu PP. Toxicity and environmental risks of nanomaterials: Challenges and future needs. Journal of Environmental Science and Health. Part C, Environmental Carcinogenesis & Ecotoxicology Reviews. 2009;**27**(1):1-35. DOI: 10.1080/10590500802708267

[45] Radziun EJ, Wilczyńska D, Książek I, Nowak K, Anuszewska EL, Kunicki A, Olszyna A, Ząbkowski T. Assessment of the cytotoxicity of aluminium oxide nanoparticles on selected mammalian cells. Toxicology In Vitro. 2011;**25**(8):1694-1700. DOI: 10.1016/j.tiv.2011.07.010

[46] Pochettino A, D'Atillio L, Bongiovanni B, Konjuh C, Bay ML, Stadler T, Dose-effect of the alumina nanoinsecticide "NSA" on THP-1 macrophage cell cultures. In: 2nd International Meeting on Pharmaceutical Industries (RICiFA), 22-23 Nov 2012, Rosario-Argentina. http://www.ricifa.com.ar/rosario2012.html

[47] Nadin SB. Genomic effects of the alumina nanoinsecticide "NSA" in human peripheral blood lymphocytes. Biocell. 2016;**40**(Suppl. 3):4. Available from: http://www.cricyt.edu.ar/biocell/XXXIV_Reunion_Cientifica.pdf

[48] Stadler T, Lascalea GE, Jahn GA. Nanotoxicology: Acute toxicity assessment of engineered nanostructured aluminium oxide. Acta Toxicológica Argentina. 2008;**16**(Suppl):1-27

[49] Chau C, Wu S, Yen W. The development of regulations for food nanotechnology. Trends in Food Science & Technology. 2007;**18**:269-280. DOI: 10.1016/j.tifs.2007.01.007

[50] Galembeck FT, Burgo AL, Balestrin BS, Gouveia RF, Silvaa CA, Galembeckd A. Friction, tribochemistry and triboelectricity: Recent progress and perspectives. RSC Advances. 2014;**4**:64280-64298. DOI: 10.1039/C4RA09604E

[51] Lindley KS, Rowson NA. Charging mechanism for particles prior to electrostatic separation. Magnetic and Electrical Separation Journal 1997;**8**(2):101. DOI: http://dx.doi.org/10.1155/1997/96189

[52] Edwards DK. A method for continuous determination of displacement activity in a group of flying insects. Canadian Journal of Zoology. 1960;**38**:1021-1025. DOI: https://doi.org/10.1139/z60-105

[53] Edwards DK. Electrostatic charges on insects due to contact with different substrates. Canadian Journal of Zoology. 1962;**40**:579-584. DOI: https://doi.org/10.1139/z62-051

[54] Edwards DK. Activity rithms of lepidopterous defoliators. l. Techniques for recording activity, eciosion and emergence. Canadian Journal of Zoology. 1964;**42**:923-937. DOI: https://doi.org/10.1139/z64-092

[55] Gan-Mor S, Schwartz Y, Bechar A, Eisikowitch D, Manor G. Relevance of electrostatic forces in natural and artificial pollination. Canadian Agricultural Engineering. 1995;**37**:189-195 Available from: http://csbe-scgab.ca/docs/journal/37/37_3_189_ocr.pdf

[56] Erickson EH, Buchmann SL. Electrostatics and pollination. In: Jones CE, Little RJ, editors. Handbook of Experimental Pollination Biology. New York: Scientific and Academic Editions; 1983. pp. 173-184

[57] Corbet SA, Beament J, Eisikowitch D. Are electrostatic forces involved in pollen transfer? Plant, Cell and Environment. 1982;**5**:125-129. DOI: 10.1111/1365-3040.ep11571488

[58] McGonigle DF, Jackson CW, Davidson JL. Triboelectrification of houseflies (*Musca domestica* L.) walking on synthetic dielectric surfaces. Journal of Electrostatics. 2002;**54**:167-177. DOI: https://doi.org/10.1016/S0304-3886(01)00177-2

[59] McGonigle DF, Jackson CW. Effect of surface material on electrostatic charging of houseflies (*Musca domestica* L). Pest Management Science. 2002;**58**:374-380. DOI: 10.1002/ps.463

[60] Jackson C, McGonigle D. Direct monitoring of the electrostatic charge of houseflies (*Musca domestica* L.) as they walk on a dielectric surface. Journal of Electrostatics. 2005;**63**:803-808. DOI: 10.1016/j.elstat.2005.03.075

[61] Scheie PO, Smyth T. Electrical measurements on cuticles excised from adult male *Periplaneta americana* (L.). Comparative Biochemistry and Physiology. 1967;**2**:547-571

[62] Vaknin Y, Gan-Mor S, Bechar A, Ronen B, Eisikowitch E. The role of electrostatic forces in pollination. Plant Systematics and Evolution. 2000;**222**(1/4):133-142. Available from:. DOI: https://link.springer.com/content/pdf/10.1007/BF00984099.pdf

[63] Prier KRS, Lighthart B, Bromenshenk JJ. Adsorption model of aerosolized bacterial spores (*Bacillus subtilis* variety niger) onto free-flying honey bees (hymenoptera: Apidae) and its validation. Environmental Entomology. 2001;**30**(6):1188-1194. DOI: https://doi.org/10.1603/0046-225X-30.6.1188

[64] Majerič P, Rudolf R, Anžel I. Thermodynamics of nanoparticles. TEHNIKA – Termodinamika Nanotehnologija. 2014;**4**(1):28-33

[65] Deacon JW. Fungal Biology. Oxford: Blackwell Publishing; 2006. p. 371

[66] Amiri A, Cholodowski D, Bompeix G. Adhesion and germination of waterborne and airborne conidia of *Penicillium expansum* to apple and inert surfaces. Physiological and Molecular Plant Pathology. 2005;**67**:40-48. DOI: doi.org/10.1016/j.pmpp.2005.07.003

[67] Scherge M, Gorb SN. Biological Micro- and Nanotribology. Berlin, Heidelberg, New York: Springer; 2001. 304 pp. DOI: 10.1007/978-3-662-04431-5

[68] Doss RP, Potter SW, Chastagner GA, Cristian JK. Adhesion of nongerminated *Botrytis cinerea* conidia to several substrata. Applied and Environmental Microbiology. 1993;**59**(6):1786-1791

[69] Mercure EW, Leite B, Nicholson RL. Adhesion of ungerminated conidia of *Colletotrichum graminicola* to artificial hydrophobic surfaces. Physiological and Molecular Plant Pathology. 1994;**45**:421-440. DOI: 10.1016/S0885-5765(05)80040-2

[70] Nechitailo G, Gordeev A. Effect of artificial electric fields on plants grown under micro-gravity conditions. Advances in Space Research. 2001;**28**(4):629-631. DOI: 10.1016/S0273-1177(01)00370-2

[71] Kuang W, Nelson SO. Low-frequency dielectric properties of biological tissues: A review with some new insights. American Society of Agricultural Engineers. 1998;**41**(1):173-184. DOI: http://dx.doi.org/10.13031/2013.17142

[72] Rezaei-Zarchi S, Imani S, Mehrjerdi HA, Mohebbifar MR. The effect of electric field on the germination and growth of *Medicago sativa* plantet, as a native Iranian alfalfa seed. Acta Agriculturae Serbica. 2012;**17**(34):105-115

[73] Koch K, Barthlott W. Superhydrophobic and superhydrophilic plant surfaces: An inspiration for biomimetic materials. Philosophical Transactions of the Royal Society A: Mathematical, Physical and Engineering Science. 2009;**367**(1893):1487-1509. DOI: 10.1098/rsta.2009.0022

[74] Fernández V, Sancho-Knapik D, Guzmán P, Peguero-Pina JJ, Gil L, Karabourniotis G. Wettability, polarity and water absorption of holm oak leaves: Effect of leaf side and age. Plant Physiology. 2014;**166**:168-180. DOI: 10.1104/pp.114.242040

[75] Savia M, Zhou Q, Koivo HN. Simulating adhesion forces between arbitrarily shaped objects in micro/nano-handling operations. In: International Conference on Intelligent Robots and Systems. Sendai, Japan. 2004. pp. 1722-1727. DOI: 10.1109/IROS.2004.1389644

[76] Knapp HF, Tami A, Brunner M, Stemmer A. Hydrophobic surface coatings on tools used for handling of micro-particles. In: 4th International Conference on Motion and Vibration Control. Zürich. 1998

[77] Greason WD. Investigation of a test methodology for triboelectrification. Journal of Electrostatics. 2000;**49**(3):344-351. DOI: 10.1109/EOSESD.1999.819082

[78] Brown RC. Tutorial review: Simultaneous measurement of particle size and particle charge. Journal of Aerosol Science. 1997;**98**:1373-1391. DOI: https://doi.org/10.1016/S0021-8502(97)00034-7

[79] Diaz AF, Felix-Navarro RM. A semi-quantitative tribo-electric series for polymeric materials: The influence of chemical structure and properties. Journal of Electrostatics. 2004;**62**:277-290. DOI: https://doi.org/10.1016/j.elstat.2004.05.005

[80] Perry RH, Green DW. Perry's Chemical Engineers' Handbook. 8th ed. London: McGraw-Hill; 2007. 2640 p

[81] Chiu SF. Toxicity studies of so-called "inert" materials with the bean weevil, *Acanthoscelidies obtectus* (Say). Journal of Economic Entomology. 1939;**32**:240-248

[82] Chiu SF. Toxicity studies of so-called "inert" materials with the Rice weevil and the granary weevil. Journal of Economic Entomology. 1939;**32**(6):810-821

[83] Alexander P, Kitchener JA, Briscoe HVA. Inert dust insecticides: Part I. Mechanism of action. Annals of Applied Biology. 1944;**3**:143-149. DOI: 10.1111/j.1744-7348.1944.tb06225.x

[84] Ebeling W. Physiochemical mechanism for the removal of insect wax by means of finely divided powders. Hilgardia. 1961;**30**:531-564. DOI: 10.3733/hilg.v30n18p531

[85] Dubinin M. The Potential Theory of Adsorption of Gases and Vapors for Adsorbents with Energetically Nonuniform Surfaces. Moscow, U.S.S.R.: Institute of Physical Chemistry, Academy of Science; 1959:235-241. DOI: http://pubs.acs.org/doi/pdf/10.1021/cr60204a006

[86] Castellanos A. The relationship between attractive interparticle forces and bulk behaviour in dry and uncharged fine powders. Advances in Physics. 2005;**54**(4):263-376. DOI: 10.1080/17461390500402657

[87] Zimon A. Adhesion of Dust and Powder. 2nd ed. New York: Springer; 1982. DOI: 10.1007/978-1-4615-8576-3

[88] Bolis V, Fubini B, Marchese L, Martra G, Costa D. Hydrophilic and hydrophobie sites on dehydrated crystalline and *Amorphous silicas*. Chemical Society, Faraday Transactions. 1991;**87**:497-505. DOI: 10.1039/FT9918700497

[89] Wang X, Xia Y. Bottom-up and top-down approaches to the synthesis of Monodispersed spherical colloids of low melting-point metals. Nano Letters. 2004;**4**(10):2047-2050. DOI: http://pubs.acs.org/doi/pdf/10.1021/nl048689j

[90] de Paula Santos F, de Campos E, Costa M, Melo FCL, Honda RY, Mota RP. Superficial modifications in TiO_2 and Al_2O_3 ceramics. Materials Research 2003;**6**(3):353-367. DOI: http://dx.doi.org/10.1590/S1516-14392003000300009

[91] Zeng XM, Martin GP, Marriott C. Particulate Interactions in Dry Powder Formulations for Inhalation. London & New York: Taylor & Francis; 2001

[92] Korunić Z. Diatomaceous Earths – Natural Insecticides. Pesticides and Phytomedicine. 2013;**28**(2):77-95. DOI: 10.2298/PIF1302077K

[93] Tsai WT, Lai CW, Hsien KJ. Characterization and adsorption properties of diatomaceous earth modified by hydrofluoric acid etching. Journal of Colloid and Interface Science. 2006;**297**:749-754. DOI: 10.1016/j.jcis.2005.10.058

[94] Mimani T, Patil C. Solution combustion synthesis of nanoscale oxides and their composites. Materials Physics and Mechanics. 2001;**4**:134-137 Available from: http://citeseerx.ist.psu.edu/viewdoc/download?doi=10.1.1.624.392&rep=rep1&type=pdf

Chapter 6

Propesticides and Their Implications

Shaon Kumar Das and Irani Mukherjee

Additional information is available at the end of the chapter

http://dx.doi.org/10.5772/intechopen.71532

Abstract

With increasing knowledge of the biochemistry and genetics of major pest insects, weeds, and agricultural pathogens, the design of such selectivity becomes a part of pesticide development and is achieved by appropriate structural modification of the parent lead molecule which is called as propesticide. In a strict sense, a propesticide is a biologically inactive compound requiring structural transformation(s) after application to become pesticidally active. Various pesticides have come to the limelight of being a propesticide by carrying out studies on their metabolic fate in organisms. Studies on the metabolic fate of diafenthiuron in vitro by liver microsomes from various vertebrates revealed a variety of possible transformations of the thiourea. Few have been developed by reversibly masking the active ingredients. Fluorinated N-acylaziridine behaves as a propesticide of the fluorinated carboxylate and the hydrolysis of the former to 2-methylaziridine and carboxylate being activation pathway. Imidacloprid and the thiazolylmethyl analogue masked with oxodioxolyl group decomposed with half life of 15.4 and 11.4 h in alkaline and physiological salt solutions, respectively, releasing imidacloprid quantitatively. New propesticide with two effects of both benzoylphenyl ureas and carbamates were designed and synthesized.

Keywords: propesticide, biologically inactive, lead molecules, oxodioxolyl, xenobiotics, selective agrochemicals

1. Introduction

Pesticide as an input in agriculture has seen changes at different stages in very few decades. Pesticides as such according to FAO may be defined as any substance or mixture of substances intended for preventing, destroying or controlling any pest, including vectors of human or animal disease, unwanted species of plants or animals causing harm during or otherwise interfering with the production, processing, storage, transport or marketing of food, agricultural commodities, wood and wood products or animal feedstuffs, or substances which may be administered to animals for the control of insects, arachnids or other pests in or on their bodies [1]. The term includes substances intended for use as a plant growth regulator, defoliant, desiccant or agent for thinning fruit or preventing the premature fall of fruit, and substances applied to crops either before or after harvest to protect the commodity from deterioration during storage and transport.

Present day pesticides which are common and widely used have number of adverse affects on non-target organisms. The consequences lead to environment toxicity affecting ecosystem [2]. Therefore safer insecticide deserves attention. A propesticide is a biologically inactive compound requiring structural transformation(s) after application to become pesticidally active. Activation process for propesticides can be one of or a combination of the three following types: (a) chemical (nonenzymatic); (b) biochemical (enzymatic); or (c) physical, e.g., photochemical [3]. In practice, however, propesticidal substances are sometimes active without chemical modification when measured *in vitro*; nevertheless, their metabolites contribute significantly to the overall biological activity of the applied material.

2. Safer pesticides

Safer pesticides are those chemicals which have no or minimum acute or chronic toxicity to mammals and harmless to non-target organisms as well as non-persistent in the environment. The harmful effects of pesticides to non-target organisms can be overcome to certain extend by increasing the selectivity pf pesticide. Pesticide selectivity as such can be attained by either physical or biochemical means that is often combined in practice. In the former, only the target species is exposed to the control agent, and this can be accomplished by special formulation or precise application techniques [4]. In the later, selectivity is based on differences of the biochemical processes or target receptors of the pest and non-pest species. Furthermore, physiochemical factor such as differential uptake by and translocation within target and non-target organisms can contribute to disparate biological activities [5]. Nevertheless, differences in metabolic pathways, which convert toxic xenobiotics into less harmful and readily excretable product, and in metabolic rates of various organisms are frequently the basis of selectivity.

3. Propesticides

A biologically inactive compound requires structural transformation(s) after application to become pesticidally active. Derivatives of known active ingredients that are converted to parent compound for activity. These are various process require for the activation of propesticide to the pesticidally active one [6]. Ideally these activation processes takes place only in the target organism. Even though it can also takes place in the environment, including soil and atmosphere. These can be one of or combination of the following four types:

3.1. Activation by primary biochemical target

This activation process is carried out in presence of certain enzyme or enzyme system. Enzymatic conversion of a proinsecticide to the active toxophore at the target tissue results in disruption of activating enzyme [7]. The activating enzymes, present in the tissue of target organism, are carrying out certain biochemical function in the body of target organism. When active toxophore are released from the proinsecticide their normal functions are discontinued resulting in killing of organisms. This is also known as suicide inactivation as their natural process is being inactivated [8]. Likewise, there also takes place the following:

a. Disruption of secondary enzyme system in the same tissue.

b. Disruption of enzyme at other target tissue.

3.2. Activation by detoxification system

By detoxification system we meant the various processes carried out by the target organisms to detoxify any xenobiotics compound [9]. These are usually degradation processes and it includes oxidation, reduction, hydrolysis or conjugation reactions. Here, the xenobiotic compounds, i.e., propesticides are acted upon by these processes resulting in production of a more toxic material than the original one.

3.3. Activation by symbiont metabolism

Most of the insects harbor symbiotic or parasitic microorganisms in their guts or hemolymph which posses enzymes lacking in them [10]. The various endogenous xenobiotic processes can be brought under control to activate proinsecticidal agents in such organism specific fashion.

3.4. Activation by symbiotic routes

In this path enzymes are not involved. Here, the activation process is not metabolic but results in toxicity because a change in the propesticide occurs in biological milieu.

4. Classification of propesticides

Propesticides can be classified in two ways like based on the number of activation steps involved and based on the type of pest to control [11]. Based on the number of activation steps involved propesticides are classified as single step activation and multiple step activation.

4.1. Single step activation

4.1.1. Juvenogens

- The term juvenogen is used to indicate a new class of the complex chemical compounds which generate products with juvenile hormone activity in response to certain biotic or environmental factors.

- Juvenogen esters when topically applied, the wax-like ester enters the insect body where it is enzymatically hydrolyzed by the carboxylesterase enzymes.
- About two orders of magnitude faster hydrolysis of the juvenogen substrate has been found in the larvae of *Dysdercus*. Here juvenogen has much higher juvenile hormone activity than the hydrolysis alcohol product itself.

4.1.2. Procarbamates and proformamidines

These are derivatives of toxic methyl carbamates insecticides which can be activated to active toxophore either by enzymatic or nonenzymatic. There are also other two possible mechanism of activation of these groups of compounds. These are:

- Acid catalyst hydrolysis of the N-S bond
- Thiol induced thiolysis to form a mixed disulfide and toxic methyl carbamate.

4.1.3. Photoactivated compounds

- Propheromones, xanthene dyes and natural photosensitizers are some types of compounds those functions by photoactivation.
- Xanthene dyes like erythrisin, fluorescin act as phototoxic agents against bacteria and insects.
- Natural photosensitizers like terthienyl and polyacetylines, DNA-damaging agents like dictamine, harmaline are some compounds which act as insecticides.

4.2. Multistep activation

Propesticides requiring more than one metabolic process for activation are known as prepropestides. These are:

4.2.1. Flourocitrate precursors

i. These occur as the toxic principle in legume genera.

ii. One of the first poisons for which the biochemical mode of action was precisely described.

iii. Nissol and Fluemethyl are two commercial product used as acaricides are of relatively lower mammalian toxicity.

4.2.2. Cycloprate

This is activated by two stage of activation process. First is hydrolysis of the cycloprate to free acid followed by formation of carnitine ester [12, 13]. Thus it inhibits the activity of carnitine in transport of fatty acids. Here, the carnitine is an amino acids commonly occurring in the liver and in skeletal muscles that function in the transport of fatty acids across mitochondrial membrane.

4.2.3. *Flouromevalenate*

These are potent inhibitors of juvenile hormone production in insects.

5. Based on the type of pest to control

Accordingly propesticides are classified as proinsecticides, proherbicides, profungicides and prorodenticides.

5.1. Proinsecticides

N-Methylcarbamates are another major group of insecticides inhibiting AChE. Although the proinsecticidal features of OP compounds were discovered after their development, proinsecticidal carbamates were designed in Fukuto's laboratory by systematic derivatization to *N*-phosphoryl, *N*-sulfenyl, and related carbamates [14–16]. The biological and toxicological properties of these carbamates could be tailored according to particular use requirements by changing the derivatizing moiety, and thus the physicochemical properties, such as lipophilicity (log P), of the resulting product. The propesticide is activated in the insect by chemical hydrolysis by thiols or other nucleophiles. Nereistoxin is a cyclic disulfide isolated from a marine annelid [17, 18]. It served as the lead compound for the development of the proinsecticides cartap and thiocyclam, both converted into dithiolane acting at the nicotinic acetylcholine receptor of insects. The precocenes, such as precocene 2, on the other hand, were isolated from *Ageratum* sp. plants and found to inhibit the terminal (oxidative) step of JH biosynthesis in the corpora allata, causing precocious development of the insect larva. These anti-juvenile hormones, also called proallatotoxins, are "suicide inhibitors" because the cytochrome P_{450} catalyzed oxidation of the chromene generates epoxide that reacts with neighboring nucleophiles of the enzyme protein, causing massive cellular damage. Diafenthiuron is a thiourea insecticide inhibiting mitochondrial ATPase and acts via its carbodiimide metabolite [19]. The phenylpyrazole fipronil contains a sulfoxide group that can undergo cytochrome P450-catalyzed oxidation in insects to yield a more potent sulfone metabolite. These are meant for controlling insect pest [20]. Some of the proinsecticides along with their active metabolite and activation processes are given in the **Table 1**.

5.2. Rodenticides precursors

Fluoroacetic acid and fluoroacetamide are "lethal precursors" to 2-fluorocitrate. Bitter scilliroside, from the red squill, can be hydrolyzed by glycosidases *in vivo* to scillirosidin, its

Propesticides	Active metabolite	Activation process
Zn_3P_2	PH_3	Hydrolysis/acidolysis
Flouroacetic acid or flouroacetamide	Flourocitric acid	Condensation with oxaloacetate/hydrolysis and condensation with oxaloacetate
Scilliroside	Scillirosidin	Hydrolysis
Bromethalin	—	*N*-Dimethylation

Propesticides	Active metabolite	Activation process
Parathion	Paraoxon	Oxidative disulfuration
Malathion	Malaoxon	Oxidative disulfuration
Disulfoton	Oxydisulfoton	Oxidation
Trichlorfon	Dichlorvos	Rearrangement/dehydrochlorination
Acephate	Methamidophos	Hydrolysis
Carbosulfan	Carbofuran	Hydrolysis
Furathiocarb	Carbofuran	Hydrolysis
Benfuracarb	Carbofuran	Hydrolysis
Thiodicarb	Methomyl	Hydrolysis
Cartap	Nereistoxin	Hydrolysis
Bensultap	Nereistoxin	Hydrolysis/
Thiocyclam	Nereistoxin	Sulfur extrusion/cyclization
Diafenthiuron	–	Oxidative desulfuration
Cycloporate	Cyclopropanecarboxylic acid	Hydrolysis
Chlorfenapyr	–	Oxidation
Sulfluramide	–	Hydrolysis
Fipronil	Fipronil sulfon	Oxidation
Tralomethrin	Deltamethrin	Debromination

Table 1. Active metabolites of proinsecticides and activation processes.

aglycone, which was suggested to be the ultimate rat toxicant. There are a few rodenticides that have either been designed to act as prorodenticides or were found to act as such.

5.3. Profungicides

Profungicides is thiram, or tetramethylthiuram disulfide that is reduced to the corresponding dithiocarbamate, the actual bioactive principle. Dithiocarbamate derivatives of glycerol and other polyols releasing or other related fungicides have also been prepared. The carbonyl group was shown to be reduced stereoselectively into the more potent fungicide triadimenol in fungi and plants, as well as in bacteria. Spirolactone derivatives of the benzoquinone fungicide chloranil provided photostable derivatives that release the parent compound by slow hydrolysis. These are meant for controlling pathogens causing plant diseases. Some of the profungicides along with their active metabolites and activation processes are given in **Table 2**.

5.4. Proherbicides

MCPB and related homologous aryloxyalkanoic acids with an odd number of CH_2 groups provide aryloxyacetic acids, such as (2-methyl-4-chlorophenoxy) acetic acid, whereas

Propesticide	Active metabolite	Activation process
Thiram	*N,N*-dimethylthiocarbamate	Reduction
Dinobuton	Dinoseb	Hydrolysis
Benomyl	Carbendazim and butyl isocyanate	Elimination/hydrolysis
Thiophanate-methyl	Carbendazim	Hydrolysis/cyclization
Triadimefon	Triadimenol	Reduction
Bupirimate	Ethirimol	Hydrolysis
Pyrazaphos		Hydrolysis
Probenazole	Saccharin	Hydrolysis
Acibenzolar-S-methyl	CGA 210007	Hydrolysis

Table 2. Active metabolites of profungicides and activation processes.

Propesticides	Active metabolite	Activation process
MCPB	MCPA	β-oxidation
Naproanilide	2-Naphthoxyacetic acid	Hydrolysis
Chlorazine	Trietazine	Dealkalization
Trietazine	Simazine	Dealkalization
EPTC	EPTC sulfoxide	Oxidation
Triallate	Triallate sulfoxide	Oxidation
Diuron	DCPMU	Oxidative dealkylation
Linuron	DCPMU	Demethoxylation
Methazole	DCPMU	Hydrolysis/reduction
Chlorthiamid	Dichlobenil	Dehydrosulfuration
Metflurazone	Norflurazon	Oxidative dealkylation
Flamprop-methyl	Flamprop	Hydrolysis
Bilanafos	Phosphinothricin	Hydrolysis
2,4-DEP	2,4-D	Hydrolysis/oxidation
Cinmethylin	2-Hydroxy-1,4-cineole	Oxidative dealkylation
Pyrazolynate	Destosyl pyrazolinate	Hydrolysis
Pyridate	CL9673	Hydrolysis
Ethephon	Ethylene	Elimination

Table 3. Active metabolites of proherbicides and activation processes.

those with an even number are degraded to herbicidally inactive phenols [21]. The occurrence of resistance in weeds to triallate has been attributed to reduced sulfoxidation, i.e., bioactivation, rates. The photosynthesis inhibitor *N,N*-dimethyl phenylurea diuron is converted into the corresponding *N*-methyl phenylurea DCPMU upon oxidative phosphorylation [22]. Dealkylation of the *N,N,N*_,*N*_-tetraethyl triazine derivative chlorazine to trietazine then to simazine increases the photosynthesis inhibitory activity by several orders of magnitude. For the rice herbicide thiobencarb (*S*-4-chlorobenzyl diethylthiocarbamate), reductive dehalogenation occurring in soil yields the *S*-benzyl derivative, believed to be responsible for phytotoxicity *in vivo*. These are meant for controlling weeds. Some of the proherbicides along with their active metabolites and activation processes are given in the **Table 3**.

6. Environmental utility of propesticides

Although commonly used structural modifications, carried out during routine structure-activity relationship studies and lead structure optimizations, affect both pharmacokinetics and pharmacodynamics, chemical alterations used in propesticide design are aimed to improve the biological profile by optimizing exclusively the pharmacokinetics of the toxicant. Potential advantages of a propesticide can be summarized as follows:

i. *Alteration of physicochemical properties.*

Altered physicochemical properties leading to improved stability, solubility, or lipophilicity influencing distribution in organism (systemicity).

ii. *Sustaining activity*

Delayed or sustained action due to the slow release of the active agent from its derivatives.

iii. *Increases selectivity*

Increased selectivity, that is, decrease toxicity toward non-target species, due to different metabolism of the parent compound and its derivatives.

7. Conclusion

The structural and metabolic diversity of various pest control chemicals shown above demonstrates the usefulness of the "Trojan horse" principle of chemical formulation. Future research efforts, either capitalizing on known pesticide design practices or discovering new ones based on differences between the xenobiotic metabolisms of various organisms, will lead to new and selective agrochemicals that find and hit the target enzyme or receptor of the pest as "magic bullets."

http://dx.doi.org/10.5772/intechopen.71532

Author details

Shaon Kumar Das[1*] and Irani Mukherjee[2]

*Address all correspondence to: shaon.iari@gmail.com

1 ICAR-National Organic farming Research Institute, Tadong, Gangtok, India

2 Division of Agricultural Chemicals, IARI, New Delhi, India

References

[1] Das SK. Recent development and future of botanical pesticides in India. Popular Kheti. 2014;**2**:93-99

[2] Kayser H, Eilinger P. Metabolism of diafenthiuron by microsomal oxidation: Procide activation and inactivation as mechanisms contributing to selectivity. Pest Management Science. 2002;**57**:975-980

[3] Das SK, Mukherjee I. Effect of light and pH on persistence of flubendiamide. Bulletin of Environmental Contamination and Toxicology. 2011;**87**:292-296

[4] Ohno I. Proinsecticide candidates *N*-(5-methyl-2-oxo-1,3-dioxol-4-yl)methyl derivatives of imidacloprid and 1-chlorothiazolylmethyl-2-nitroimino-imidazolidine. Bioorganic and Medical Chemistry Letter. 2007;**17**(16):4500-4503

[5] Das SK, Mukherjee I. Flubendiamide transport through packed soil columns. Bulletin of Environmental Contamination and Toxicology. 2012;**88**:229-233

[6] Chen L, Wang Q. Inseticidal benzoylphenylurea-s-carbamate: A new propesticide with two effects of both benzoylphenylureas and carbamates. Journal of Agriculture and Food Chemistry. 2007;**53**:38-41

[7] Das SK, Avasthe RK, Singh R, Babu S. Biochar as carbon negative in carbon credit under changing climate. Current Science. 2014;**107**:1090-1091

[8] Das SK, Mukherjee I. Influence of microbial community on degradation of flubendiamide in two Indian soils. Environmental Monitoring and Assessment. 2014;**186**:3213-3219

[9] Das SK, Mukherjee I. Effect of moisture and organic manure on persistence of flubendiamide in soil. Bulletin of Environmental Contamination and Toxicology. 2012;**88**:515-520

[10] Das SK. Role of micronutrient in rice cultivation and management strategy in organic agriculture—A reappraisal. Agricultural Sciences. 2014;**5**:65-769. DOI: 10.4236/as.2014.59080

[11] Das SK, Avasthe RK, Gopi R. Vermiwash: Use in organic agriculture for improved crop production. Popular Kheti. 2014;**2**:45-46

[12] Masaki H, Shinijiro N. New selective grass herbicides and their hydrolytic properties as pro-herbicides. Bioscience, Biotechnology, and Biochemistry. 1992;**56**:620-623

[13] Das SK, Mukherjee I, Das SK. Metsulfuron-methyl herbicide on dehydrogenase and acid phosphatase enzyme activity on three different soils. International Journal of Bio-Resource & Stress Management. 2017;**8**(2):236-241

[14] Das SK. Mode of action of pesticides and the novel trends—A critical review. International Research Journal of Agricultural Science and Soil Science. 2013;**3**(11):393-401

[15] Meazza G, Rama F, Bettarani F. Synthesis and bioactivity of some fluorine-containing benzoyl arylureas. Part I: Insecticidal-acaricidal products in which the aryl groups bears a triflouromethyl-substituted alkyl or alkenyl side chain. Pesticide Science. 1992;**35**:137

[16] Mate CH, Mukherjee I, Das SK. Persistence of spiromesifen in soil: Influence of moisture, light, pH and organic amendment. Environmental Monitoring and Assessment. 2015;**187**:7. DOI: 10.1007/s10661-014-4207-6

[17] Okada H, Koyanagi T, Yamada N. Synthesis and antitumor of prodrugs of benzoylphenylureas. Chemical and Pharmaceutical Bulletin. 1994;**42**:57-61

[18] Das SK, Roy A, Barman H. Fungi toxic efficiency of some plant volatile essential oils against plant pathogenic fungi. African Journal of Microbiology Research. 2016;**10**(37): 1581-1585

[19] Das SK. Scope and relevance of using pesticide mixtures in crop protection: A critical review. International Journal of Environmental Science and Toxicology. 2014;**2**(5):119-123

[20] Hamm S, Cherton L, Menguy R, Delorme, Louveaux A. Potential proinsecticides of flourinated carboxylic acids. III. Evaluation of the N-acylaziridine structure by ^{19}F NMR monitoring of the in vitro behaviour in insect tissues. New Journal of Chemistry. 1999;**23**:1239-1244

[21] Das SK, Mukherjee I, Das SK. Dissipation of flubendiamide in/on Okra [*Abelmoschus esculenta* (L.) Moench] fruits. Bulletin of Environmental Contamination and Toxicology. 2012;**88**:381-384

[22] Roy A, Das SK, Tripathi AK, Singh NU. Biodiversity in North East India and their Conservation. Progressive Agriculture. 2015;**15**(2):182-189

Chapter 7

Toxic Effects of the Organophosphorus Insecticide Fenthion on Growth and Chlorophyll Production Activity of Unicellular Marine Microalgae *Tetraselmis suecica*: Comparison between Observed and Predicted Endpoint Toxicity Data

Maria C. Vagi, Andreas S. Petsas, Maria D. Pavlaki,
Niki M. Smaragdaki and Maria N. Kostopoulou

Additional information is available at the end of the chapter

http://dx.doi.org/10.5772/intechopen.72321

Abstract

This chapter provides the results of a laboratory ecotoxicological study conducted to assess the acute toxicity of the organophosphorus pesticide fenthion toward the marine microalgal species *Tetraselmis suecica*. Bioassays were performed, and algal densities and chlorophyll pigments fractions were measured in the exponential phase after 96 h of exposure to fenthion. Two quantitative structure activity relationships (QSARs) were used to estimate the toxicity of 13 primary metabolites and degradation products of fenthion toward the selected organism; the first was based on the use of the n-octanol/water partition coefficient, whereas the second was based on the solubility of the compound in water. Results revealed that fenthion can have marked effects on the growth and photosynthesis of the target primary producers of marine ecosystems *T. suecica*. The parent pesticide toxicant was found not toxic to the tested algal species up to 1.00 mg L^{-1}, while higher treatment concentrations not only affected algal densities and significantly decreased specific growth rate values (μ) ($p < 0.05$) but also decreased the contents of photosynthetic pigments. The comparison between the observed and the predicted toxicity values of the parent compound fenthion indicated that the predictive capability of the QSARs applied can be considered highly satisfactory. Consequently, both QSAR models were used for the prediction of toxicity data of fenthion's principal metabolites and degradation products.

Keywords: fenthion, *Tetraselmis suecica*, toxicity test, ecotoxicology, pigment biomarker, QSARs

IntechOpen

1. Introduction

Fenthion (*O,O*-dimethyl *O*-(4-(methylthio)-*m*-tolyl) phosphorothioate) is a contact and stomach systemic organophosphorus pesticide, used as a wide-spectrum insecticide for numerous crops against many suckings and biting pests. It was developed in 1960 and first commercialized by Bayer Agriculture in the USA as an insecticide/acaricide for mosquito and insect control that is commercially available worldwide in several formulations [1].

In Greece, only one formulation of fenthion was registered by Bayer CropScience Hellas, with the trade name LEBAYCID 50% EC (containing 50% w/v fenthion as the active ingredient), which is classified as dangerous for humans, but is not classified for aquatic organisms [2]. This insecticide is extremely effective in controlling the major insects infecting olives, such as the olive fruit fly *Bactrocera oleae (Dacus)*, the olive kernel borer or olive moth *Prays oleae*, the black scale *Saissetia oleae (Olivier)*, and *Margaronia* sp. and other masticatory insects. Although *B. oleae* is considered the most serious insect, all aforementioned insects are widely distributed in the Mediterranean region and occur on olives at population densities causing important economic losses. Therefore, fenthion was for many years one of the most commonly used pesticides in Greek territory and in the Mediterranean area generally [2].

The available information on the production and use of pesticides in general and hence of organophosphates as well is limited, fragmentary and in some cases unreliable [3]. On the basis of the limited information received from the Mediterranean countries, fenthion was one of the important compounds used during the 1980s and 1990s among other organophosphorus pesticides [3]. According to data provided by the Greek Ministry of Rural Development and Food, it appears that the quantities of fenthion that were used for agricultural purposes during the years 1983, 1984, 1985, 1986, 1987, 1988, and 1989 in Greece were 216,892; 409,139; 24,359; 197,843; 87,787; 160,433; and 213,514 tons of active ingredient, respectively [3].

Since June 2007, fenthion is no longer approved by the Greek Ministry of Rural Development and Food because of an excess number of poisoning-related events and ecotoxicology effects on nontarget organisms (Greek Ministry Decision, Register Number 122914–27/4/2005, 2005), apart from its 120 days of exceptional authorization (from May 1, 2009 to August 31, 2009) in accordance with Art. 8(4) of Directive 91/414/EEC for the treatment of olive trees against *Dacus oleae* (Greek Ministry Decision, Register Number 128569–11/5/2009c IN, 2009).

Although fenthion was developed as a safe pesticide because it is not easily converted to the possibly highly toxic oxon derivative (fenthion oxon) in mammalian species, however according to relative literature, many of its metabolites were detected in various plants, animals, and environmental matrices [4–7]. Kitamura et al. demonstrated that the in vivo metabolism of fenthion in fish leads to the formation of two metabolites, fenthion sulfoxide and fenthion oxon [4], while other studies proved that fenthion and its oxidation products were accumulated in fish [5]. Oxidation products of fenthion, including fenthion oxon, were also detected in house mosquitoes exposed to fenthion [6]. It has also been reported that fenthion was converted to fenthion oxon in the aqueous environmental bodies [7]. On the contrary, the toxicity and the metabolism of this organophosphorus insecticide have not been extensively studied in aquatic microspecies, such as microalgae.

Microalgae are important inhabitants of aquatic ecosystems, where they fulfill critical roles in primary productivity, nutrient cycling, and decomposition. Detrimental effects of pesticides on algae may have subsequent impacts on higher trophic levels [8]. It has been well established that changes in the macromolecular composition of phytoplankton species or shifts in community composition can affect the growth rate of zooplankton grazers [9]. Unquestionably, aquatic environments receive direct and indirect pesticide inputs, inevitably exposing microorganisms to pesticides. Millions of pounds of active pesticide ingredients are applied in coastal watersheds each year, and in addition, pesticides may affect marine inhabitants via spills, runoff, and drift [10].

Toxicity data involving ecotoxicology of fenthion toward nontarget microorganisms are limited. Most studies have focused on microbial degradation and biotransformation of fenthion rather than impacts on natural microbial populations and communities. Furthermore, studies of fenthion effects on soil microbes are far more common than studies of toxicology assessments in aquatic environments. Published data regarding marine or estuarine microorganisms are even scarcer [11].

The aims of the present survey were (i) to assess the acute toxicity of the organophosphorus insecticide fenthion toward nontarget aquatic microorganisms, such as marine algae, (ii) to investigate the possibility of using the parameter of chlorophyll pigments as biomarkers of exposure to fenthion, (iii) to compare the observed and predicted endpoint toxicity data and evaluate the predictive capability of two QSARs based on physicochemical properties of target organic toxicant (n-octanol/water partition coefficient and water solubility), and (iv) to predict the toxicity of 13 principal metabolites and degradation products of fenthion toward the selected marine microalgae.

2. Materials and methods

2.1. Organism and culture conditions

Tetraselmis is a genus of a marine, motile, green phytoplankton (Prasinophyceae) that has very high lipid levels and also contains natural amino acids that stimulate feeding by other marine animals [12]. For this reason, it is used as a food source for feeding marine crustaceans, especially shrimp and mollusks. *Tetraselmis suecica* (Kylin) Butcher [12] (formerly known as *Platymonas suecica*, by Kylin, [13]) is a free-living, flagellate species that was initially isolated from seawater of the English and Swedish coasts, but later research has suggested that it is probably cosmopolitan [14]. The unicellular marine microalga *T. suecica* that is used in the bioassays of the present study is a strain of phytoplankton that is commonly cultivated in shellfish husbandries [15] and has been routinely cultivated by our laboratory [16]. This species was chosen because it is easy to be cultivated [15], and its response in toxicity tests is highly reproducible [16].

Unialgal cultures of the species were maintained in liquid *f*/2 growth medium as recommended by Provasoli-Guillard National Center for Culture of Marine Phytoplankton (CCMP) [17]. More specifically, the selected strain of microalga was cultured in natural seawater, which had been filtered through a 0.45 μm Millipore filter, autoclaved at 121°C for 20 min, and enriched with several nutrients, trace metals, and vitamins. The final concentration of each component in *f*/2 growth medium was $NaN0_3$, 0.882 mM; $NaH_2PO_4 * H_2O$, 36.2 μM; $N\alpha_2SiO_3 * 9H_2O$, 0.106 mM; $FeCl_3 * 6H_2O$, 11.7 μM; $Na_2EDTA * 2H_2O$, 11.7 μM; $CuSO_4 * 5H_2O$, 39.3 nM; $Na_2MoO_4 * 2H_2O$, 26.0 nM; $ZnSO_4 * 7H_2O$, 76.5 nM; $CoCl_2 * 6H_2O$, 42.0 nM; $MnCl_2 * 4H_2O$, 0.910 μM; thiamine HCl (vit. B_1), 0.296 μM; biotin (vit. H), 2.05 nM; and cyanocobalamin (vit. B_{12}), 0.369 nM. Stock solutions of these components were autoclave-sterilized except in the case of vitamins which were filter-sterilized by passing through a Millipore filter (0.45 μm) before supplementing to the growth medium. Salinity of the seawater was 35.0 ± 0.1%, and the initial pH of the cultures was 8.0 ± 0.1 [18].

One hundred milliliters of inoculated growth medium *f*/2 with *T. suecica* at a cell density of 1×10^5 cells mL^{-1} was contained in 250 mL flasks with air-permeable stoppers. The cultures were incubated under continuous illumination with cool-white fluorescent lights emitting a radiant energy of 4300 Lux equivalent to 12.9 W m^{-2}. Temperature was maintained stable in a temperature-controlled growth chamber (Snijders Scientific B.V., The Netherlands), at 20.0 ± 0.3°C. The test vessels containing the cultures during the course of the experiments were gently shaken by hand once per day in order to keep the cells in free suspension, to facilitate CO_2 mass transfer from air to water, and in turn to reduce pH shift. Hence, variations in pH during the 96 h of incubation were within the limit of ±1.0 unit. All glassware and mediums used were previously sterilized by autoclaving at 121°C for 20 min, and all handlings were made under aseptic conditions so as to avoid contamination from bacteria or other species of algae [16].

2.2. Test chemicals, reagents, and standards

The tested compound fenthion was an analytical grade (purity >99.5%), obtained from Dr. Ehrenstorfer-Schäfers (Augsburg, Germany) and used without further purification. Pure fenthion is a colorless, almost odorless liquid, while technical product of fenthion (95–98% pure) is a brown oily liquid with a weak garlic odor. Data for other physiochemical properties of fenthion, taken from reference [19], include melting point, 7°C; boiling point, 87°C at 0.01 mmHg; vapor pressure, 1.4 mPa at 25°C; water solubility, 55 mg L^{-1} (at 20°C and pH = 7); log $K_{ow,}$ 4.84; and M_r, 278.34. **Figure 1** shows the chemical structure of the target compound.

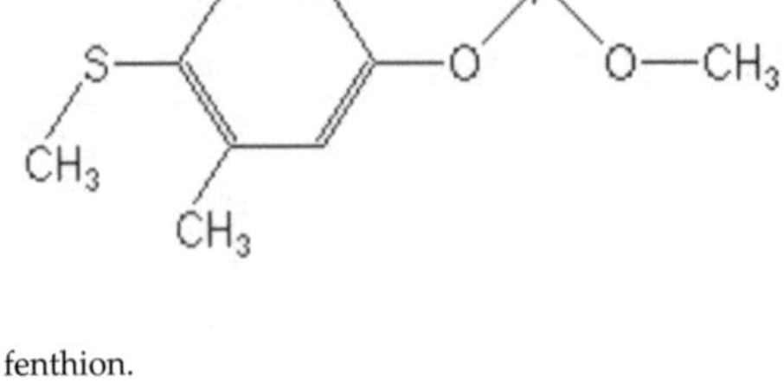

Figure 1. Chemical structure of fenthion.

Due to low water solubility of the tested substance, acetone was used for the preparation of its stock solutions. Hence, acetone was used as the carrier solvent of the compound to the bioassays, since previous experiments proved that this solvent up to a final concentration of 0.5 μL mL^{-1} in *f*/2 medium did not affect the growth rate of the tested algae [16]. Stock standard solutions of fenthion (1000 and 10,000 mg L^{-1}) were prepared by dissolving the required amounts in acetone (HPLC grade) and were stored under refrigeration.

Pesticide-grade organic solvents such as acetone, hexane, methanol, and dichloromethane were purchased from Pestiscan (Labscan Ltd., Ireland). Organic-free water was prepared with a Milli-Q/Milli-Ro system (Millipore Corp., Bedford, USA). Other chemical reagents and solvents used were of HPLC grade and procured from Merck (Merck, Germany).

2.3. Procedure for the study of the stability of fenthion

The stability of fenthion in seawater was determined under the experimental conditions employed for the incubation of the cultures. Therefore, parallel experiments were performed without algae using all six test concentrations that were chosen for the toxicity treatments (0.5, 1.0, 1.5, 2.0, 2.5, and 3.0 mg L^{-1}) and prepared in 0.45 μm GF/F-filtered natural seawater. Triplicate samples were prepared in 250 mL borosilicate flasks, and each contained 100 mL of pesticide solution. At various time intervals (0, 24, 48, 72, and 96 h, as for the toxicity tests), 5 mL aliquots of the aqueous reaction solutions were withdrawn for analysis.

Fenthion concentrations were quantitatively confirmed by gas chromatography analysis after liquid–liquid extraction of the fortified aqueous samples with hexane (2 × 5 mL). Organic extracts were dried over anhydrous sodium sulfate, and 1 μL of the extracts was injected into gas chromatographic system (Hewlett-Packard 5890) equipped with a nitrogen-phosphorus detector (GC-NPD). A 30 m × 0.32 mm i.d. × 0.25 film thickness fused silica-bonded phase capillary column (MDN-5, Supelco, USA) was used for the chromatographic separation of target analyte and oven temperature programmed at 150°C for 3 min; increased from 150–170°C at 20°C min^{-1}; then increased from 170–190°C at 2°C min^{-1}; after that increased from 190–250°C at 15°C min^{-1}; and was held to 250°C for 15 min. Helium was used as the carrier gas at constant flow of 1.2 mL min^{-1} during GC analysis. Injection technique was on column. Detector's temperature was 280°C, while hydrogen and air were used as NPD's airs with flows of 3.5 and 110 mL min^{-1}, respectively.

2.4. Acute toxicity test and pesticide treatment

Bioassays were performed according to the OECD Guideline 201 for testing the effects of chemicals on alga growth inhibition test [18], with some modifications. Cells in the exponential phase of growth were collected from stock cultures (called as pre-cultures and incubated under the previously mentioned conditions) and for this reason used as the inoculum. The initial algal density in each one of the experimental treatments was of 1×10^5 cells mL^{-1} [16].

The experimental design and test conditions were identical for all replicates performed. Each chemical bioassay included the below-described treatments: a control (C) containing no pesticide; a control containing acetone as carrier solvent of the organic toxicants, in concentration

0.05% (C + A); and various toxicant exposure concentrations of fenthion (in mg L^{-1}), following the results of preliminary range-finding experiments conducted for the tested compound previously [16]. Algae were exposed to the concentration series of 0.50, 1.00, 1.50, 2.00, 2.50, and 3.00 mg L^{-1} of fenthion, respectively. Each treatment contained three replicate flasks. The environmental conditions during the experiments were the same as the growth conditions stated in paragraph 2.1.

Cell densities were assessed daily (after 0 h, 24 h, 72 h, and 96 h of incubation) by microscope counting using an improved Neubauer hemocytometer. Lugol solution was added to the samples (ratio of lugol/culture media, 1/10 v/v) to prevent the natural movement of *T. suecica* cells. Specific growth rate (μ), which is the proportional rate of increase in cell density per unit of time, was calculated for each treatment sample according to Eq. (1):

$$\mu = \frac{\ln N_t - \ln N_o}{t - t_o} \tag{1}$$

where t_o is the time of test start ($t_o = 0$ d) and t is the time of test termination ($t = 4$ d), while N_o and N_t are the initial and final cell densities at times t_o and t, respectively [20]. Inhibition of algal growth as a reduction in specific growth rate, relative to control cultures grown under identical conditions (%*I*), was estimated from the relationship (2):

$$\%\text{Inhibition} = \%I = \frac{\mu_{control} - \mu_{pesticide}}{\mu_{control}} \times 100 \tag{2}$$

where $\mu_{pesticide}$ is the algal growth rate in the presence of the tested compound and $\mu_{control}$ is the growth rate in the untreated control. The EC_{20} and EC_{50} values (pesticide concentration required to cause a 20 and 50% reduction in growth, respectively) were calculated using linear regression analysis of transformed pesticide concentration as logarithm data versus percentage inhibition. Furthermore, the no-observed-effect concentration (NOEC) was defined as the highest tested concentration of fenthion below which no reduction in reproduction was observed after a 96-h exposure period, while the lowest-observed-effect concentration (LOEC) was defined as the lowest tested concentration of fenthion at which reduction of algal growth was observed after a 96-h exposure period. The maximum acceptable toxicant concentration (MATC) was defined as a hypothetical threshold concentration that was calculated as the geometric mean between NOEC and LOEC concentrations.

The contents of acetone-soluble chlorophyll pigments, chlorophyll-a (Chl_a), chlorophyll-b (Chl_b), and chlorophyll-c (Chl_c), contained in 10 mL of culture medium at the end of incubation (96 h), were determined according to the spectrophotometric method described in detail by Strickland and Parsons [21].

2.5. Prediction of toxicity values of primary metabolites and degradation products of fenthion: Quantitative structure activity relationships (QSARs)

Two structure-toxicity relationships have been proposed by Vagi [22] for the growth inhibition of the marine microalga *T. suecica* exposed to various organophosphates such as dimethoate, parathion methyl, parathion ethyl, and its oxidative metabolite paraoxon ethyl. These are described by Eqs. 3 and 4:

$$\log(1/EC_{50}) = 0.5415 \log P_{OW} - 2.6499, \text{ with correlation coefficient } R^2 = 0.9689 \quad (3)$$

$$\log(1/EC_{50}) = -0.6367 \log S + 0.5338, \text{ with correlation coefficient } R^2 = 0.9094 \quad (4)$$

where P_{OW} is the n-octanol/water partition coefficient and S is the solubility of the compound in water at 20°C in mg L^{-1}. P_{OW} characterizes the lipophilicity of the molecule and quantifies its tendency to partition between water and suspended solids, its partitioning and uptake into biota (bio-concentration) as well as its adsorption to sediments; thus, log P_{OW} is considered to be a parameter describing the kinetics of uptake of chemicals from water. On the contrary, S value encodes quantitative information on the hydrophilicity of the compound and comprises the inclination of the chemical to remain into the aqueous phase.

Since the experimental determination of log P_{OW} and S values can be impractical and time-consuming, accurate and straightforward methods for the determination of this important property are available. As it is well known, computational chemical methods, which only require the chemical structure of the molecule, are one of the most famous and useful approaches to estimate several physicochemical properties such as log P_{OW} and S values. In the absence of a complete set of reliable experimental values and in order to obtain homogeneous values, the hydrophobicity and hydrophilicity of degradation products of parent compound fenthion, expressed as log P_{OW} and S values, respectively, were calculated according to the available scientific prediction methods provided by Virtual Computational Chemistry Laboratory (VCCLAB) by using the ALOGPS 2.1 ++ logP/logS calculation software program [23].

2.6. Data reliability and statistical analysis

Independent experiments were repeated three times, and each sample (treatment and/or control culture) was repeated three times. Mean values ± standard deviations (SD) are shown in the figures, and tables are presented in this chapter. Data collected were calculated as percentages, and arcsine was transformed (arcsine $\sqrt{x}$) and analyzed using one-way analysis of variance (ANOVA). Variances were considered equal ($p > 0.05$) based on Kolmogorov–Smirnov test for homogeneity of variance. The highest concentration of toxicant demonstrating no effect as compared to the controls was estimated by Dunnett's test for statistical significance ($p > 0.05$) with SPSS software program.

3. Results

3.1. Fenthion stability

Experimental data of the present study concerning the stability of fenthion in 0.45 μm GF/F-filtered seawater during 96 h of exposure to illumination of cool-white fluorescent lights (4300 Lux or 12.9 W m^{-2}) and at 20.0 ± 0.3°C are summarized in **Table 1**.

Treatment concentration (mg L^{-1})	Remaining pesticide (%) over time				
	0 h	24 h	48 h	72 h	96 h
0.50	98.88	96.64	95.74	92.76	91.95
1.00	98.55	97.17	95.81	94.47	93.14
1.50	97.48	96.94	95.54	93.33	93.25
2.00	98.23	97.65	95.09	94.87	92.79
2.50	100.76	97.75	96.02	95.18	93.62
3.00	102.98	99.36	97.45	94.63	92.89

Values are the percentage (%) of the nominal concentration remaining and represent the means of triplicate tests.

Table 1. Fenthion loss in 0.45 μm GF/F-filtered seawater during 96 h of illumination of cool-white fluorescent lights (4300 Lux or 12.9 W m^{-2}) and at 20.0 ± 0.3°C.

It must be mentioned that the loss of the target organophosphorus toxicant from solutions over 96 h was similar in all three replicates (data not shown), and the mean values of triplicates are presented. It is obvious that the initial test concentrations (0 h) were between 97.48 and 102.98% of the nominal concentrations, while after 96 h of exposure (under 4300 Lux and at 20 ± 0.3°C), concentration of fenthion in seawater had reduced to between 91.95 and 93.62% of the nominal concentrations. This percentage of loss (less than 10%) is in accordance with the condition set by the OECD for the validity of the test that requires no more than 20% of the test chemical to be lost during the conducted toxicity test [18].

3.2. Toxicity of fenthion on growth of marine alga *T. suecica*

Algae exhibit several responses to toxicants, including growth inhibition and stimulation and morphological and physiological changes [24]. During the performance of present bioassays, the algal cells changed morphologically when treated with fenthion and observed under the optical microscope. Changes in cell shape, color, and size were observed as some cells became darker in color and in other cases were swollen as well. Moreover, many cell divisions were found abnormal because when the material cell divided, the descendant cells remained attached and the daughter cells were not separated. Thus, usual algal aggregations could also be observed. The above phenomenon indicated that fenthion could have been a potential of mutagenic effects on *T. suecica*. This was consistent with the phenomena observed by Li et al. while studying the effect of the synthetic pyrethroid cypermethrin on *Scenedesmus obliquus* [25], but to our knowledge, it has not been reported as a toxic effect of any other organophosphorus pesticide on algal toxicity tests.

Cultured in different concentrations of fenthion, the algal growth curves of T. suecica are shown in **Figure 2(a–f)**. The results contained in these charts indicated that cells in fenthion-treated medium grew slower than those in control group.

Furthermore, the mean specific growth rates (μ) of target alga when exposed to the range of concentrations of the tested toxicant for the three replicated bioassays conducted are summarized

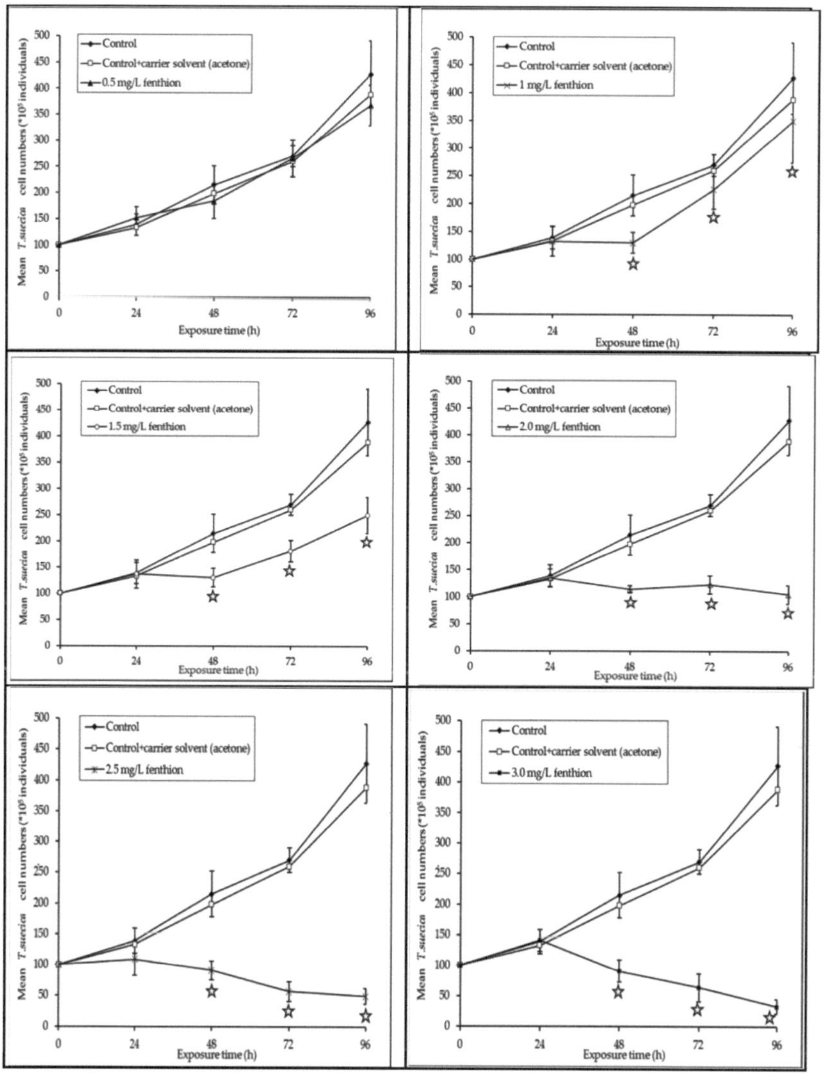

Figure 2. Effects of different concentrations of fenthion on growth of *Tetraselmis suecica*. [error bars represent standard deviations of three replicates, ☆significantly different as compared to the controls ($p < 0.05$)].

in **Table 2**. As it can be seen in these results, acetone controls, containing the carrier solvent, did not differ significantly from blank controls. On the contrary, it is observed that the organophosphorus toxicant consistently inhibited the algal population growth in concentrations from 0.50 to 3.00 mg L^{-1}, and the specific growth rate became remarkably lower with the increase of fenthion concentration, which demonstrated that fenthion can inhibit growth of *T. suecica* at the concentration range tested. Obtained values of μ indicated that significant inhibition of the algal densities occurred in treatment levels of fenthion above 1.00 mg L^{-1}. Pesticide concentrations of 1.50, 2.00, 2.50, and 3.00 mg L^{-1} significantly reduced ($p < 0.05$) *T. suecica* densities after 96 h of exposure. Severe reduction in growth occurred at concentration 2.00 mg L^{-1}, while 2.50 and 3.00 mg L^{-1} were found to be lethal.

Treatment concentration		μ^a (d^{-1})
(mg L^{-1})	(μmol L^{-1})	
Control	Control	0.58 ± 0.04
Control + carrier solvent (acetone)	Control + carrier solvent (acetone)	0.54 ± 0.04
0.50	1.80	0.37 ± 0.03
1.00	3.59	0.34 ± 0.01
1.50	5.39	0.33 ± 0.02*
2.00	7.19	0.31 ± 0.03*
2.50	8.98	0.23 ± 0.04*
3.00	10.78	0.01 ± 0.04*

[a]Based on algal densities.
*Significantly different as compared to the controls ($p < 0.05$).
Values are means ± standard deviation of three replicates.

Table 2. Specific growth rate (μ) of *T. suecica* after 96 h of exposure to treatments of fenthion.

Similar results were found by other authors who tested the influence of the organophosphorus insecticide fenitrothion on *Nannochloris oculata* and reported that treatment concentrations higher than 1.00 mg L^{-1} affected algal growth, whereas μ values decreased significantly by concentrations 5.00, 10.00, and 15.00 mg L^{-1} [26].

Using the toxicity data contained in **Table 2**, an estimate of NOEC and LOEC values would be 1.00 and 1.50 mg L^{-1}, respectively, while MATC calculated as the geometric mean between the NOEC and LOEC was estimated to be 1.22 mg L^{-1}. The experimental results of the present study confirmed that fenthion is slightly less toxic toward the target marine microalgae than it was previously reported as the values of NOEC, LOEC, and MATC were reported to be 0.50, 1.00, and 0.70 mg L^{-1}, respectively [16]. The percentage of inhibition data relative to growth in untreated controls (%*I*) was calculated according to Eq. (2).

Figure 3a shows the concentration-response curve of fenthion to *T. suecica,* which obviously corresponded to typical and characteristic sigmoid form (S-shape). Obtained toxicity values of %*I* were linearly related to transformed pesticide concentration values by logarithmic conversion (logC), and the plotted log transformation of the "concentration effect" line is presented in **Figure 3b**. The S-shape is again evident, but the curve approaches a straight line, and a linear portion of the curve is obvious and presented in **Figure 3c**.

The linear regression equation that was derived from this linear part of the curve is described by Eq. (5):

$$\%I = 181.26\ \log C + 17.01, \text{ with correlation coefficient } R^2 = 0.9778 \quad (5)$$

where %*I* represents the percentage inhibition ($0 \le \%I \le 100$) and C is the pesticide concentration (in mg L^{-1}). High value of correlation coefficient showed that data fitted satisfactorily to the linear model.

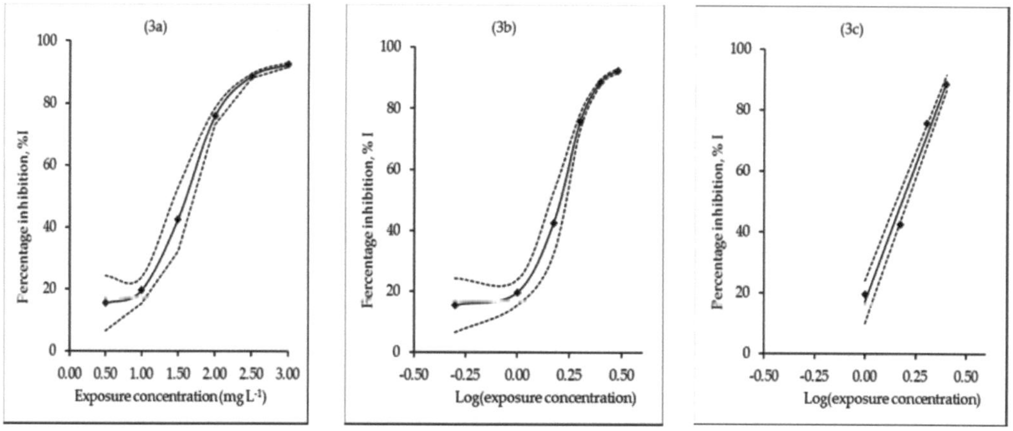

Figure 3. Concentration-response curves of fenthion on growth of *Tetraselmis suecica*. (a) Percentage inhibition versus concentration. (b) Percentage inhibition versus logarithm transformation of concentration. (c) Linear portion of percentage inhibition versus logarithm transformation of concentration. (Dotted lines on each side of the curve represent the 95% confidence limits).

Acute toxicity values of EC_{20} and EC_{50} (in mg L^{-1}) at 96 h were obtained by the above-described relationship (5), and the calculated data are presented in **Table 3**.

EC_{50} values of target compound estimated in the present work are in accordance with toxicity data reported in the literature for the same toxicant toward other green algal species, such as *Scenedesmus subspicatus* (EC_{50}, 1.79 mg L^{-1}) [19] and *Kirchneria subcapitata* (EC_{50}, 1.1 mg L^{-1}) [27]. A large number of ecological toxicity data of fenthion toward several nontarget aquatic organisms are available in published literature. Reported toxicity values indicated that fenthion is moderately toxic to estuarine and marine fish (LC_{50} for *Cyprinodon variegatus*, 1200 μg L^{-1} [27]), moderately highly toxic to freshwater fish on an acute basis (LC_{50} for *Oncorhynchus mykiss*, 0.83 mg L^{-1} and for *Lepomis macrochirus*, 1.7 mg L^{-1} [27]), very highly toxic to estuarine and marine invertebrates (LC_{50} or EC_{50} for *Crassostrea virginica*, 321 μg L^{-1} and for *Americamysis bahia*, 0.22 μg L^{-1} [27]), very highly toxic to freshwater invertebrates on an acute basis (EC_{50} for *Daphnia magna*, 5.2 μg L^{-1} [27]), and finally moderately toxic to nontarget aquatic plants such as marine and freshwater diatoms (EC_{50} for *Skeletonema costatum*, 0.4 mg L^{-1} and for *Navicula pelliculosa*, 1.0 mg L^{-1} [27]).

3.3. Toxicity of fenthion on chlorophyll pigment production of marine algae *T. suecica*

T. suecica Kylin (Butch) is an algal strain that its mass and biochemical composition, mainly in protein, chlorophyll-a, and RNA content, have shown great variabilities, which are related

EC_{20} 96 h		**EC_{50} 96 h**	
(mg L^{-1})	**(mol L^{-1})**	**(mg L^{-1})**	**(mol L^{-1})**
1.04	3.74×10^{-6}	1.52	5.46×10^{-6}

Table 3. Acute toxicity values of fenthion to *T. suecica* after 96 h of exposure.

to changes in nutrient concentrations and that phenomenon has a marked effect on the nutritive value of this microalga as feed in marine culture. According to relevant literature, these observed changes in the chlorophyll-a level either in the stationary or in logarithmic phase of growth were related to nitrogen depletion [15].

Fenthion belongs to a chemical group of pesticides called organophosphates, which share a common mechanism of toxicity; they all affect the nervous system by inhibiting acetylcholinesterase (AChE). The physiological role of AChE is the cleavage of the neurotransmitter acetylcholine at cholinergic synapses and neuromuscular junctions, thereby terminating the neurotransmitter's effects on the postsynaptic membrane. The toxicity of insecticidal organophosphates also called the anticholinesterase insecticides (anti-ChEs) is based on their inhibition of AChE, which results in interference with proper neurotransmission. Therefore, fenthion is not expected to be a direct inhibitor of pigment synthesis nor to induce a direct oxidative stress as a consequence of its biochemical mode of action that would destroy chlorophyll pigments. However, pigment content may change in response to the cascade of events following contamination with the pesticide, regardless of its different mode of action [28, 29]. Results of the effect of fenthion on the pigments were expressed either as pigment content of the culture or as percentage inhibition of pigment increase. These results are shown in **Table 4** and **Figure 4**.

From the collected experimental data, it became obvious that fenthion decreased the contents of photosynthetic pigments (Chl_a, Chl_b, Chl_c, and Chl_{tot}) and statistically significantly different as compared to the controls occurred in photosynthetic activity of *T. suecica* cells that were treated with concentration of fenthion above 1.00 mg L^{-1}. Values of concentration ratio of chlorophyll-a/chlorophyll-b (Chl_a/Chl_b) were calculated, and results are shown in **Table 4**.

Treatment level (mg L^{-1})	**Chl_a (μg L^{-1})**	**Chl_a (pg $cell^{-1}$)**	**Chl_b (μg L^{-1})**	**Chl_b (pg $cell^{-1}$)**	**Chl_c (μg L^{-1})**	**Chl_c (pg $cell^{-1}$)**	**Chl_{tot} (μg L^{-1})**	**Chl_{tot} (pg $cell^{-1}$)**	**Chl_a/Chl_b**
Control	1125.169	2.632	463.586	1.084	140.194	0.3279	1728.949	4.044	2.43
Control + acetone	1263.114	3.249	540.690	1.391	100.796	0.2593	1904.600	4.899	2.34
0.50	995.208	2.704	378.972	1.030	93.088	0.2530	1467.268	3.987	2.63
1.00	1106.970	3.163	378.506	1.081	95.916	0.2740	1581.392	4.518	2.92
1.50	888.978*	3.556*	296.504	1.186	86.216	0.3449	1271.698*	5.087*	3.00
2.00	587.190*	5.592*	192.943*	1.838*	77.249*	0.7357*	857.383*	8.166*	3.04
2.50	424.708*	8.668*	143.219*	2.923*	37.895*	0.7734*	605.822*	12.364*	2.97
3.00	307.121*	9.307*	105.281*	3.190*	30.632*	0.9282*	443.034*	13.425*	2.92

Mean values of three replicates.

*Statistically significantly different as compared to the controls ($p < 0.05$).

Table 4. Effect on the pigment content of *T. suecica* after 96 h of exposure to treatments of fenthion.

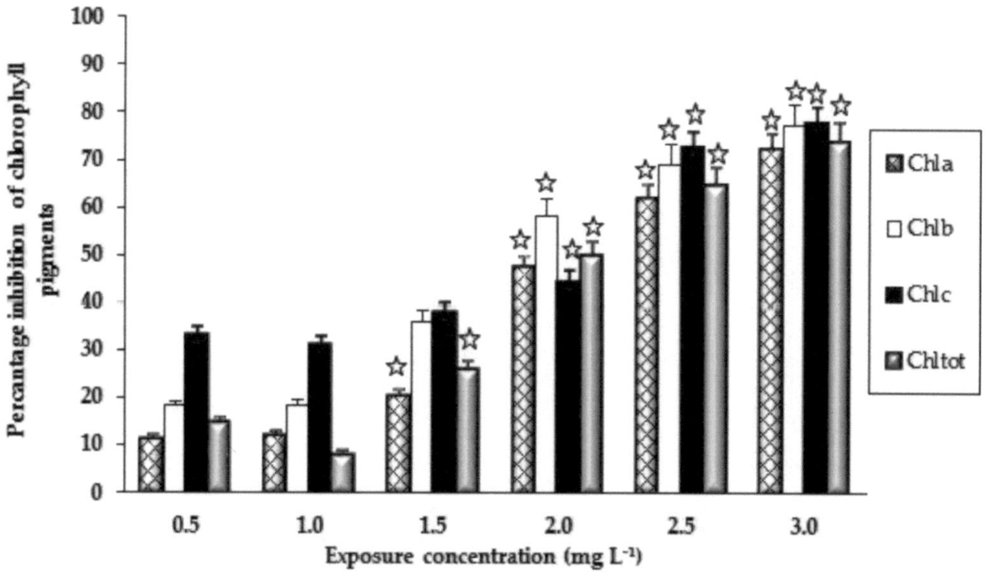

Figure 4. Effect of fenthion on the photosynthetic activity of *Tetraselmis suecica*. [Error bars represent standard deviations of three replicates, ☆ significantly different as compared to the controls ($p < 0.05$)].

It was clear that while cell density decreased with increasing exposure treatments of fenthion, values of Chl_a/Chl_b ratio remained stable or increased, suggesting that the biomass of algae was affected by the organophosphorus insecticide much more strongly than the structure of the chlorophyll body. These data were in agreement with those of Li et al., who reported that cypermethrin induced a drastic decrease in the growth and photosynthesis of *Scenedesmus obliquus* and that production of each chlorophyll pigment separately was more sensitive to cypermethrin than the ratio of Chl_a/Chl_b [25].

Linear correlations between cell density and chlorophyll pigment concentrations of chlorophyll-a, chlorophyll-b, chlorophyll-c, and total chlorophyll were calculated and are described by Eqs. (6)–(9), respectively:

$$Chl_a = 0.00002\,N + 312.41466, \text{ with correlation coefficient } R^2 = 0.9459 \tag{6}$$

$$Chl_b = 0.00001\,N + 84.68639, \text{ with correlation coefficient } R^2 = 0.9213 \tag{7}$$

$$Chl_c = 0.000002\,N + 33.815385, \text{ with correlation coefficient } R^2 = 0.8322 \tag{8}$$

$$\text{and } Chl_{tot} = 0.00003\,N + 430.91643, \text{ with correlation coefficient } R^2 = 0.9530 \tag{9}$$

where Chl_a, Chl_b, and Chl_c are the concentrations of chlorophyll pigments in culture media, Chl_{tot} is the sum of Chl_a, Chl_b, and Chl_c (all in μg L^{-1}), and N is the cell number (in cells).

Compound	Parameters		Predicted EC_{50} 96 h (mg L^{-1})	
	log P_{ow}	log S (at 20°C) (mg L^{-1})	QSAR-log P_{ow} [Eq. (3)]	QSAR-log S [Eq. (4)]
Fenthion	3.73	7.49	4.27	1.35
Fenthion sulfoxide (I)	2.18	240.00	29.47	0.92
Fenthion sulfone (II)	2.34	44.86	24.14	1.09
Fenthion oxon (III)	2.30	810.00	25.38	0.83
Fenthion oxon sulfoxide (IV)	0.87	2597.00	150.94	0.75
Fenthion oxon sulfone (V)	0.91	1773.00	143.59	0.78
Demethyl fenthion (VI)	3.07	93.78	9.72	1.01
Demethyl fenthion sulfoxide (VII)	1.59	1650.00	61.51	0.78
Demethyl fenthion sulfone (VIII)	1.67	590.00	55.67	0.85
Demethyl fenthion oxon (IX)	1.81	2540.00	46.75	0.75
Demethyl fenthion oxon sulfoxide (X)	0.42	19140.00	264.53	0.65
Fenthion phenol (XI)	2.49	1067.00	20.02	0.81
Fenthion phenol sulfoxide (XII)	1.19	8533.00	101.28	0.69
Fenthion phenol sulfone (XIII)	1.04	3163.00	122.11	0.74

Table 5. Calculated EC_{50} values for *T. suecica*.

The above-described linear correlations resulted in high values of correlation coefficients ($R^2 > 0.8322$), a fact which indicated that the use of chlorophyll measurements to estimate biomass concentration is reliable and validated the possibility of using cell chlorophyll content to assess the state of the cells after 96 h of exposure to fenthion, as previously described by other authors for other cases of bioassays [30, 31]. These results confirmed that the commonly accepted hypothesis of chlorophyll pigment content being proportional to growth rate of microalgal species [32] applies for toxicity assessment of fenthion on marine phytoplanktonic species such as *T. suecica*.

Acquired values of chlorophyll content expressed in pg. $cell^{-1}$ are summarized for each pigment in **Table 4**. Unfortunately, there is lack of available published information on photosynthetic activity of this species, and the few data are restricted only to chlorophyll-a concentrations [15]. It is observed that when incubated with fenthion concentrations equal or below 1.50 mg L^{-1}, the content of chlorophyll-a/cell of *T. suecica* reached values between 2.632 and 3.556 pg. $cell^{-1}$. Similar results were obtained by other authors who reported values of chlorophyll-a/cell between 3.1 and 3.8 pg./cell [15]. On the contrary, after exposure to treatment levels higher than 1.50 mg L^{-1}, an increase in these values occurred, and the

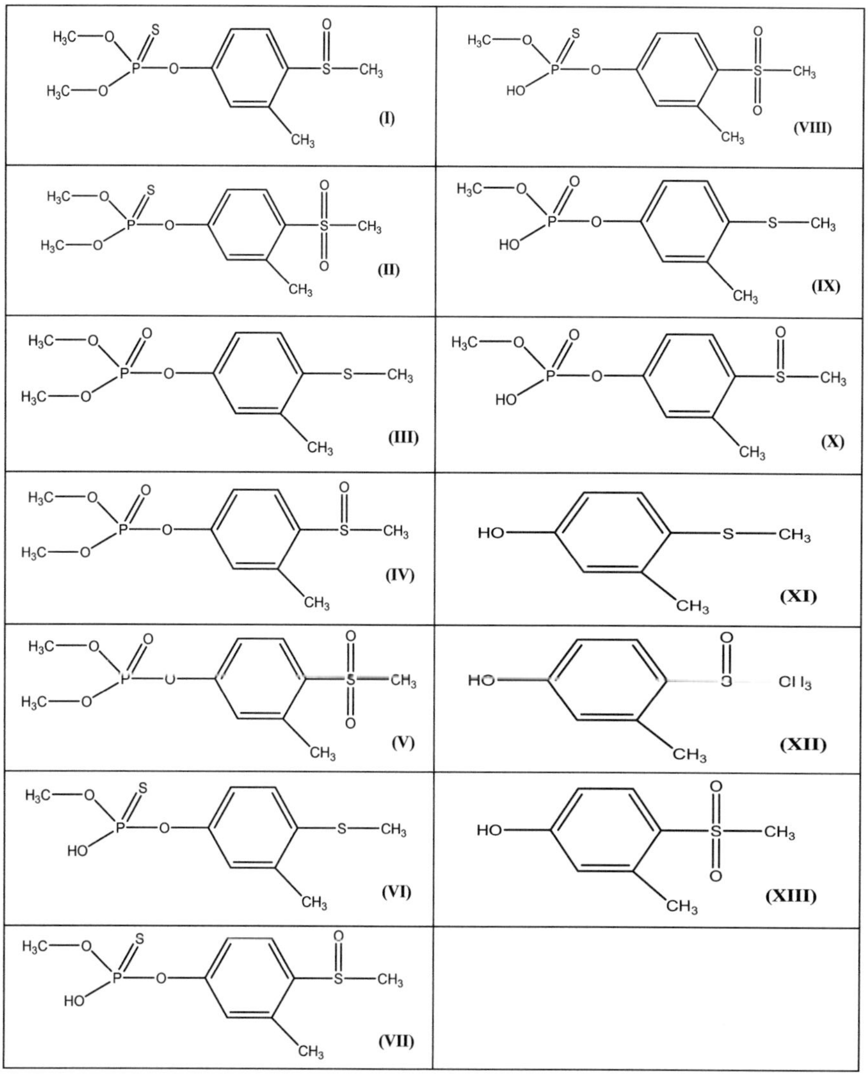

Figure 5. Principal metabolites and degradation products of fenthion. [(I) Fenthion sulfoxide, (II) fenthion sulfone, (III) fenthion oxon, (IV) fenthion oxon sulfoxide, (V) fenthion oxon sulfone, (VI) demethyl fenthion, (VII) demethyl fenthion sulfoxide, (VIII) demethyl fenthion sulfone, (IX) demethyl fenthion oxon, (X) demethyl fenthion oxon sulfoxide, (XI) fenthion phenol, (XII) fenthion phenol sulfoxide, and (XIII) fenthion phenol sulfone].

range was from 5.592 to 9.307 pg. cell^{-1}. This phenomenon suggested that either the determination of chlorophyll-a concerned pigment amounts that were extracted from dead cells as well as from the live ones or that under the stress due to high exposure levels of fenthion, a mechanism of stimulation occurred by the incubated strain and resulted in the increase of chlorophyll-a concentration.

3.4. Toxicity of the metabolites of fenthion on growth of marine alga *T. suecica*

The abiotic and biotic degradation of organophosphorus pesticides has been extensively studied in a large number of studies. Various data concerning the metabolism of several organophosphates in terrestrial and aquatic species, either in vivo or in vitro, are available [4–6]. After the application of Eqs. (3) and (4) for the prediction of the toxicity of 13 principal metabolites and degradation products of fenthion that have been identified in environmental samples, the predicted EC50 values for *T. suecica* are listed in **Table 5**. Additionally, the chemical structures of those compounds, called as metabolites and degradation products, are presented in **Figure 5**.

According to predicted EC_{50} values of Eq. (3), the parent chemical was more toxic than all of its metabolites, while on the contrary, according to Eq. (4), all of the 13 metabolites and degradation products of fenthion were expected to be more toxic than the parent compound. The acquired toxicity based on QSAR containing log P_{OW} data [Eq. (3)] followed the order: fenthion > demethyl fenthion > fenthion phenol > fenthion sulfone > fenthion oxon > fenthion sulfoxide > demethyl fenthion oxon > demethyl fenthion sulfone > demethyl fenthion sulfoxide > fenthion phenol sulfoxide > fenthion phenol sulfone > fenthion oxon sulfone > fenthion oxon sulfoxide > demethyl fenthion oxon sulfoxide. Interestingly, fenthion oxon that is the transformation product of fenthion by oxidative desulfuration was not predicted to be as toxic to *Tetraselmis suecica* up as the parent compound fenthion. EC_{50} value of fenthion oxon was estimated to be 25.38 mg L^{-1}, approximately six times higher than EC_{50} of fenthion, which was 4.27 mg L^{-1}. That fact could be attributed either to physicochemical properties of the compound (such as the highest water solubility and lowest octanol water partition coefficient) or to low persistence of the molecule into marine ecosystems as it undergoes under rapid hydrolysis. On the contrary, the calculated toxicity based on QSAR containing log *S* data [Eq. (4)] followed the order: demethyl fenthion oxon sulfoxide > fenthion phenol sulfoxide > fenthion phenol sulfone > fenthion oxon sulfoxide = demethyl fenthion oxon > fenthion oxon sulfone = demethyl fenthion sulfoxide > fenthion phenol > fenthion oxon > demethyl fenthion sulfone > fenthion sulfoxide > demethyl fenthion > fenthion sulfone > fenthion. This observation is in accordance with EC_{50} values found for the organophosphorus pesticide fenamiphos and its oxidation products fenamiphos sulfoxide (FSO) and fenamiphos sulfone (FSO_2) toward the aquatic alga *Pseudokirchneriella subcapitata* and the terrestrial alga *Chlorococcum* sp., which proved that parent compound was less toxic than its metabolites [33].

4. Conclusions

Based on the results of the current study, it appeared that fenthion can be highly toxic to the marine microalgal strain *T. suecica*. Experimental data revealed that the examined organophosphorus pesticide had marked effects on the growth of the tested algae since treatment concentrations above 1.00 mg L^{-1} affected algal densities and significantly decreased specific growth rate values. The finding that reduction of chlorophyll pigment production was observed due to exposure to fenthion indicated that this parameter could be used as a pollution biomarker. Moreover, two quantitative structure activity relationships, QSARs, based on physicochemical properties of the toxicants were applied for the prediction of toxicity values EC_{50} of the

http://dx.doi.org/10.5772/intechopen.72321

13 principal metabolites and degradation products of parent organic compound, fenthion. Finally, the comparison between observed and predicted endpoint toxicity data showed that the predictive capability of both employed QSARs could be considered highly satisfactory.

Author details

Maria C. Vagi[1]*, Andreas S. Petsas[1,2], Maria D. Pavlaki[3], Niki M. Smaragdaki[1] and Maria N. Kostopoulou[1]

*Address all correspondence to: mbag@env.aegean.gr

1 Department of Marine Sciences, Laboratory of Environmental Quality and Geo-spatial Applications, School of the Environment, University of the Aegean, Mytilene, Lesvos, Greece

2 Department of Food Sciences and Nutrition, School of the Environment, University of the Aegean, Myrina, Lemnos, Greece

3 Department of Biology and Center for Environmental and Marine Studies (CESAM), University of Aveiro, Aveiro, Portugal

References

[1] U.S. National Library of Medicine, TOXNET, TOXICOLOGY DATA NETWORK. Hazardous Substance Data Bank, (HSDB): Fenthion, CASRN: 55-38-9. Available from: http://toxnet.nlm.nih.gov/cgi-bin/sis/htmlgen?HSDB [Accessed: Aug 31, 2017]

[2] Tsatsakis A, Tsakiris I, Tzatzarakis M, Agourakis Z, Tutudaki M, Alegakis A. Three-year study of fenthion and dimethoate pesticides in olive oil from organic and conventional cultivation. Food Additives & Contaminants: Part A. 2003;**20**:553-559

[3] UNEP/FAO/WHO/IAEA. Assessment of the state of pollution of the Mediterranean Sea by Organophosphorus compounds. Athens, Greece: Map Technical Reports Series, No. 58; 1991

[4] Kitamura S, Suzuki T, Kadota T, Yoshida M, Ohash K, Ohta S. In vitro metabolism of fenthion and fenthion sulfoxide by liver preparations of bream, goldfish, and rats. Drug Metabolism and Disposition. 2003;**31**(2):179-186

[5] Tsuda T, Kojima M, Harada H, Nakajima A, Aoki S. Accumulation and excretion of fenthion, fenthion sulfoxide and fenthion sulfone by killifish (*Oryzias latipes*). Comparative Biochemistry and Physiology C. 1996;**113**:45-49

[6] Stone BF. Metabolism of fenthion by the southern house mosquito. Journal of Economic Entomology 1969;**62**:977-981

[7] Lacorte S, Jeanty G, Marty JL, Barceló D. Identification of fenthion and temephos and their transformation products in water by high-performance liquid chromatography with

diode array detection and atmospheric pressure chemical ionization mass spectrometric detection. Journal of Chromatography. A. 1997;**777**:99-114

[8] DeLorenzo ME, Scott GI, Ross PE. Toxicity of pesticides to aquatic microorganisms: A review. Environmental Toxicology and Chemistry. 2001;**20**:84-98

[9] Ahlegren G, Lundstedt L, Brett M, Forsberg C. Lipid composition and food quality of some freshwater phytoplankton for cladoceran zooplankters. Journal of Plankton Research. 1990;**12**:809-818

[10] Pehkonen SO, Zhang Q. The degradation of organophosphorus pesticides in natural waters: A critical review. Critical Reviews in Environmental Science and Technology. 2002; **32**(1):17-72

[11] Hanazato T, Kasai F. Effects of the organophosphorus insecticide fenthion on phyto- and zooplankton communities in experimental ponds. Environmental Pollution. 1995; **88**:293-298

[12] Butcher RW. An introductory account of the smaller algae of British coastal waters. Part I: Introduction and Chlorophyceae. Great Britain, London: Minist. Agric. Fish. Food, Fisheries Investigations. 1959;**Series IV**(Part 1):1-74

[13] Kylin H. Die Chlorophyceen der schwedischen Westküste. Lunds Universitets Arsskrift. N.F., Avd. 1949;**245**(4):1-79

[14] van den Hoek C, Mann D, Jahns HM. Algae-an Introduction to Phycology. UK: Cambridge University Press; 1995. 623 pages. ISBN: 0521304199

[15] Fabregas J, Herrero C, Cabezas B, Abalde J. Mass culture and biochemical variability of the marine microalga *Tetraselmis suecica* Kylin (butch) with high nutrient concentrations. Aquaculture. 1985;**49**:231-244

[16] Vagi MC, Kostopoulou MN, Petsas AS, Lalousi ME, Rasouli C, Lekkas TD. Toxicity of organophoshorous pesticides to the marine alga *Tetraselmis suecica*. Global NEST Journal. 2005;**7**(2):222-227. Available from: http://journal.gnest.org/sites/default/files/Journal%20 Papers/paper_9_VAGI_375.pdf [Accessed: Jan 23, 2017]

[17] Provasoli-Guillard National Center for Culture of Marine Phytoplankton (PGCCMP). Medium Recipes. Medium f/2. Bigelow Laboratory for Ocean Sciences. West Boothbay Harbor, Maine, USA. 2010. Available from: https://ncma.bigelow.org/[Accessed: Aug 31, 2017]

[18] Organization for Economic Co-Operation and Development, (OECD). OECD Guidelines for Testing of Chemicals, Test No. 201: Freshwater Alga and Cyanobacteria, Growth Inhibition Test. OECD Publications Office -Environ. Health and Safety Div., OECD Environ. Directorate, Paris, France. 2011. Available from: http://dx.doi.org/10.1787/9789264069923-en [Accessed Aug 31, 2017]

[19] Tomlin CDS. The Pesticide Manual: A World Compendium. 11th ed. Farnham, Surrey, UK: British Crop Protection Council; 1997. 1606 pages. ISBN:1901396118

[20] Nyholm N, Källqvist T. Environmental toxicology review: Methods for growth inhibition toxicity tests with freshwater algae. Environmental Toxicology and Chemistry. 1989; **8**:689-703

[21] Strickland JDH, Parsons TR. A Practical Handbook of Seawater Analysis. Bulletin 167. 2nd ed. Ottawa: Fisheries Research Board of Canada; 1972. 309 pages. Available from: https://epic.awi.de/39262/1/Strickland-Parsons_1972.pdf [Accessed: Dec 9, 2017]

[22] Vagi MC. Hydrolysis and adsorption study of selected organophosphorus pesticides in aquatic and soil systems. Evaluation of their toxicity on marine algae. [PhD thesis] (in Greek). Greece: University of the Aegean, Mylilene; 2007. 355 pages. Available from: https://www.didaktorika.gr/eadd/handle/10442/17779 [Accessed: Aug 31, 2017]

[23] Virtual Computational Chemistry Laboratory (VCCLAB). 2005. Available from: http://www.vcclab.org [Accessed: Aug 31, 2017]

[24] Lewis MA. Algae and vascular plant tests. In: Rand GM, editor. Fundamentals of Aquatic Toxicology-Effects, Environmental Fate, and Risk Assessment. 2nd ed. Boca Raton: CRC Press; 1995. pp. 135-169. ISBN: 156032-091-5

[25] Li X, Ping X, Xiumei S, Zhenbin W, Liqiang X. Toxicity of cypermethrin on growth, pigments and superoxide dismutase of *Scenedesmus obliquus*. Ecotoxicology and Environmental Safety. 2005;**60**:188-192

[26] Ferrando MD, Sancho E, Andreu-Moliner E. Chronic toxicity of fenitrothion to an alga (*Nannochloris oculata*), a rotifer (*Brachionus calyciflorus*) and the cladoceran (*Daphnia magna*). Ecotoxicology and Environmental Safety. 1996;**35**:112-120

[27] United States Environmental Protection Agency (US-EPA). Transmittal of EFED RED for the List A Chemical Fenthion, Case #0290, Chemical #053301. Washington, DC, USA: Office of Prevention, Pesticides, and Toxic Substances; 1996

[28] Petsas AS, Vagi MC. Effects on the photosynthetic activity of algae after exposure to various organic and inorganic pollutants: Review. In: Jacob-Lopes E, editor. Chlorophyll. Rijeka, Croatia: InTech; 2017. pp. 37-77. DOI: 10.5772/67991. Available from: https://www.intechopen.com/books/chlorophyll/effects-on-the-photosynthetic-activity-of-algae-after-exposure-to-variousorganic-and-inorganic-poll [Accessed: Sep 18, 2017]

[29] Couderchet M, Vernet G. Pigments as biomarkers of exposure to the vineyard herbicide flazasulfuron in freshwater algae. Ecotoxicology and Environmental Safety. 2003; **55**(3):271-277

[30] Wilhelm C, Manns L. Changes in pigmentation of phytoplankton species during growth and stationary phases-consequences for the reliability of pigment based methods of biomass determination. Journal of Applied Phycology. 1991;**3**:305-310

[31] Davidson K, Flynn KJ, Cunningham A. Relationships between photopigments, cell carbon, cell nitrogen and growth rate for a marine nanoflagellate. Journal of Experimental Marine Biology and Ecology. 1991;**153**:87-96

[32] Borowitzka LJ, Borowitzka MA. B-carotene (Provitamin A) production with algae. In: Vandamme EJ, editor. Biotechnology of Vitamins, Pigments and Growth Factors. London: Elsevier Applied Science; 1989. pp. 15-26. ISBN: 978-94-009-1111-6

[33] Cáceres T, Megharaj M, Naidu R. Toxicity and transformation of fenamiphos and its metabolites by two micro algae *Pseudokircheriella subcapitata* and *Chlorococcum* sp. Science of the Total Environment. 2008;**398**:53-59